ANIMAL WELFARE SCIENCE, HUSBANDRY AND ETHICS

ANIMAL WELFARE SCIENCE, HUSBANDRY AND ETHICS

The Evolving Story of Our Relationship with Farm Animals

Mark Fisher

Foreword by Temple Grandin

First published 2018

Copyright © Mark Fisher 2018

All rights reserved. No part of this publication may be reproduced, stored in a retrieval system, or transmitted, in any form or by any means, electronic, mechanical, photocopying, recording or otherwise, without prior permission of the copyright holder.

Published by
5M Publishing Ltd,
Benchmark House,
8 Smithy Wood Drive,
Sheffield, S35 1QN, UK
Tel: +44 (0) 1234 81 81 80
www.5mpublishing.com

A Catalogue record for this book is available from the British Library

ISBN 9781789180084

Book layout by Servis Filmsetting Ltd, Stockport, Cheshire
Printed by Hobbs The Printers Ltd, Totton, Hampshire

In memory of

Ron Kilgour (1932–1988)

and

Bully Todd (2015–2018)

Contents

Acknowledgements x
Foreword by Temple Grandin xii

Chapter One	The simplicity and complexity of animal welfare	1
Chapter Two	Drawing on the wealth of agriculture	42
Chapter Three	High farming and hard work	83
Chapter Four	Husbandry from beyond the farm gate	121
Chapter Five	People are people through animals	174
Chapter Six	Thinking like a mountain	212
Chapter Seven	The fall and rise of the hunter-gatherer	243

References 249

Acknowledgements

Generations of shepherds, ranchers, farmers, cockies, gauchos, yeomen, peasants, cowhands, slaves, herders and many others have provided the canvas, the paints, the brushstrokes and the hues for civilization. I am indebted to the past generations, to the generous hospitality of the present generation who have helped me understand our debt, and to the future generations who will nurture the relationship between ourselves and farm animals. *He aha te mea nui o te ao, he tangata he tangata he tangata – What is the most important thing in the world, it is the people, the people, the people.*

Teachers, scientists, philosophers and librarians in many countries have helped shape my professional career and thus this book. I thank Clive Dalton, Peter Fennessy, Neville Gregory, Peter Mason, David Mellor, Ruth Newberry, Barbara Nicholas, Mat Stone, along with Bernard Rollin and the late Sir Colin Spedding for their initial encouragement, for continuing this important role in science and ethics.

The incredible support of those helping source and provide some of the figures has been amazing. Thank you to all of you, especially the generosity of Louise Goossens, Michelle Isaac, Heather O'Brien, Ed Hansen, Stephen Barker, Jens Sigsgaard, Jen Burford, Lynsey Swift, Alison Clague, Jim Yurasek, James Fisher, Kim Ellett, John Ditchburn, Alistair Lawrence, Mary Foster, Allan Pearson, Glen Kilgour, Barry Friend, Paul Stancliffe, Jim Webster, Keith Fisher, Angela Singer, Ellie Hutchinson, Kathryn Harvey, Anne Maguire, Geraldine Margery, Lisa Potter Bonvicini, Ross Buscke, Kim Bush, Banksy, PCO and Fiona Fisher. Similarly, the generosity of the Capital and Coast District Health Board, Harvard Art Museums/Fogg Museum, New Dishley Society, Merial NZ (now Boehringer Ingelheim Animal Health New Zealand), Ron Kilgour Memorial Trust, University of Guelph Library, Chartered Accountants Australia New Zealand, J. Paul Getty Museum, European Commission and the Solomon R. Guggenheim Museum is much appreciated.

Excerpts from *Animal Farm* by George Orwell (Copyright © George Orwell, 1945) reprinted by permission of Bill Hamilton as the Literary Executor of the Estate of the Late Sonia Brownell Orwell; *The Magpies by Denis Glover* by permission of Denis Glover Estate and Pia Glover, the copyright holder; and *Animals and Nature* by Rod Preece © University of British Columbia Press 1999 (all rights reserved by the Publisher).

Sarah Hulbert's incredible talent in knowing what I had to do and 'letting me run with the ball' is very gratefully acknowledged. Similarly, the support of the rest of the production team at 5M Publishing, especially Alessandro Pasini, Jeremy Toynbee, Nick Morton, Drew Stanley, Geraldine Beare, Victoria Delahunty and Katherine Weber.

This book began with my mother who, in response to a farming newspaper article entitled 'Farmers Grumpy over Welfare Plan', asked 'Does ethics address who morally pays for what?' My family, in both New Zealand and the UK, and especially Marilyn and Sarah, have, over the intervening two decades, in an incredibly different and invaluable number of ways, shared, helped, inspired, supported, tolerated, nourished and sustained my efforts. You have been outstanding in everything from providing insights, to entertaining visitors, to identifying works and quotes. Thank you.

Foreword

Temple Grandin

Professor of Animal Science, Colorado State University

Dr Fisher's book is unique because it covers both the history of animal domestication and broader, complex animal welfare issues that are not covered in other books. The discussion of how animals were domesticated and how farming developed will provide valuable insight. The readers will learn about the history of many issues.

For me, it has always been obvious that animals have emotions. This may have been influenced by my study of both neuroscience and animal ethology. I learned how the brains of animals and humans have great similarities. I will never forget the anatomy class that I audited during the early 1980s, where I had the opportunity to dissect a human brain. At the time I was participating in these labs, I was looking at Golgi-stained pig brains. This was part of my dissertation on the effects of environmental enrichment on the neural development in the somatosensory cortex of the pig. When the two brains are compared, the main difference is the size of the cortex. The parts of the brain where the emotions are processed looked very similar. People have a bigger computer sitting on top of lower brain centres that are approximately the same size as they are in pigs.

It is only recently that the neuroscience of animal emotions has entered the veterinary and animal science literature. It had been in the neuroscience literature for decades. In 1997, I wrote a paper titled 'Assessment of stress during handling and transport' in the *Journal of Animal Science*. This was one of the first papers where research on the neuroscience of fear had been presented to animal scientists. Many people who worked with farm animals were reluctant to recognize that animals had emotions.

The *five freedoms* was one of the early conceptual frameworks for improving animal welfare. Preventing suffering is not sufficient, an animal should have a life worth living where it can have positive emotions. As animal welfare definitions evolved, the requirement of providing a truly good life has received more and more emphasis. However, there are still people in farming who do not want to recognize animal emotions. With my background in neuroscience, rejecting animal emotions seemed silly. Let us look at the facts: antidepressants such as Prozac (fluoxetine) work in dogs, and the brain neurotransmitters are the same. A neuroscientist, Jaak Panksepp, has clearly shown that the seven basic emotional systems are similar in animals and humans. The Panksepp emotional systems are covered in our book *Animals Make Us Human* by Grandin and Johnson. They are: FEAR, which motivates animals to escape from predators and danger, RAGE (anger), PANIC (separation distress), SEEK (the urge to explore), LUST (sex), NURTURE (mother young caring and social grooming) and PLAY.

Animals that are living in a good environment would have more positive emotions such as SEEK, LUST, NURTURE and PLAY than animals in a poor environment.

Complex issues

Another really important feature of Mark Fisher's book is that it covers broader, complex animal welfare issues. A worldwide issue is the conflict between raising farm animals on the land and co-existing with wildlife. Basic issues will have to be addressed about the use of the land. Unfortunately, many of these issues have become highly emotional, and a reasoned discussion is often difficult. Dr Fisher's book will provide vital historical information.

There is a need for different books that approach the issue of welfare from different viewpoints. In my own work, I have concentrated on practical ways that people who are purchasing food or working on farms can prevent suffering. There will always be a need for this. If I am a buyer or a supply chain manager at a large food company I have to prevent bad practices. I must also develop and manage auditing programs to ensure that welfare standards are being followed. Both farm managers and buyers can use my papers and books to implement a program. That is the major purpose of my book *Improving Animal Welfare: A Practical Approach*. Dr Fisher's book will be essential reading when broader issues have to be addressed.

I have worked with the meat industry for over 40 years to improve animal welfare at slaughter. When I first started my career, I was warned that the animals would know that they were walking up a ramp to get slaughtered. To determine if this was true, I went back and forth between the slaughter house and a feedlot where cattle were being vaccinated in a race. Cattle behavior was the same in both places. If the cattle knew they were going to be slaughtered, they should have been more agitated at the slaughter plant.

Being a visual thinker who thinks in pictures enabled me to realize that I should observe what the cattle were seeing. Small distractions such as a paper towel hanging from a dispenser or seeing motion through a crack in the door would stop cattle movement. Removing visual distractions improved cattle flow through the handling system. The animals were more concerned about visual distractions than getting slaughtered.

Mark Fisher says we should develop different ways of understanding animals. Being a visual thinker instead of a verbal word thinker helped me to understand animals. The animals we use for food should be given a good life that is worth living.

CHAPTER ONE

The simplicity and complexity of animal welfare

I am a dog. I have four legs and a tail. The food I have is meat. Sometimes I dig in the garden. This is bad. When I am good I chase sheep. (Fisher, 1961)

The wolf, the dog and mankind

A collie, longingly and with devotion and loyalty, looks across to distant hills. The bronze statue erected by the region's run holders or farmers, and those who appreciate the value of the sheepdog, is a tribute to the animal that helped them graze sheep in New Zealand's (NZ) mountainous high country. Situated near the Church of the Good Shepherd, this dog is visited by tourists from around the world, its popularity undoubtedly due, at least in part, to the especially long and close relationship between mankind and dogs (Figure 1.1).

Our worlds are the culmination of thousands of years of entwined biological, social and cultural evolution. The diversity and success of both dogs and people reflect the inextricable and synergistic relationship we have shared for at least 15,000, perhaps as long as 100,000 years (Vila et al., 1997). It is a relationship founded on symbiosis – among other things, dogs provide assistance and companionship in return for food and shelter – a relationship at the heart of being human. As many have noted, to be human is to have an innate attraction to nature, to bond with special others (family, friends, homes, land and countries) including drawing in and interacting with animals and other life forms, in diverse and complex ways. These bonds can be strong and crucial to the health and well-being of individuals and communities, and some can be very lasting.

Our worlds are the result of thousands of years of interactions, coevolution and displacement of animals and humans. Wolves and Neanderthals have given way to Dalmatians, Jack Russells and mongrels, and athletes, soldiers, teachers and many more. The diversity and success of modern

Figure 1.1 Beannachdan Air Na Cu Caorach (blessings on the sheepdogs) at Lake Tekapo, New Zealand (*Photo: Mark Fisher*). (Available in colour in the plate section between pages 178–179.)

dogs and humans are arguably due to the extraordinary inextricable and synergistic relationship we share, evident in the relative reduction in brain volume from the wolf to the dog as also occurred in humans as each used and became dependent on the other's intellect. Dogs, unlike wolves, can read humans, domestication selecting animals able to understand humans (Hare et al., 2002). Reduced fear, acceptance of the other, and even a love of travel are some of the human and canine traits responsible for this long-lasting symbiotic relationship. As the science fiction novelist, Kim Stanley Robinson (2012), remarked: 'individual intelligence probably peaked in the Upper Paleolithic, and we have been self-domesticated creatures ever since, dogs when we had been wolves'.

The diversity and success of both dogs and people are inarguable, we share a good life. Dog-wolves provided hunter-gatherers with comfort and companionship, found and attacked prey, cleaned the camp of food scraps and vermin, or were a source of food (Derr, 2012; Fagan, 2015). There were few limits to their utility as they guarded, herded, hauled, warmed beds, were sacrificed to the gods, provided medicines, and accompanied and guided humans in this world and the next. The spiritual importance of the relationship was also evident in the belief that dog-wolves rebalanced humans with the forces of their world, judged ethical behaviour, ensured rituals were carried out properly and protected humanity.

To be a dog, as with other species having a symbiotic relationship with humans, is to survive and thrive – there are fewer canines in wild or natural populations. Many of the close ties between dog-wolves and man, forged some 15,000 or more years ago, remain today. Like many animals, dogs can be sources of labour, entertainment, learning and commerce, used to maintain social order and provide companionship, used symbolically in expressing human characters and relationships, or be part of the wild. However, it is as members of our households and families that dogs have claimed a part in our lives. At one level, they were workers, for example providing the labour for a kitchen's rotisserie. The Turnspete or turnspit dog was bred to run on a wheel or turnspit to turn meat whilst it was roasting over an open fire. These spaniel-sized, short-legged and long-bodied dogs were considered ugly, but their strong legs and stamina ensured meat was cooked evenly. Popular in the 16th to 19th centuries in large kitchens in England and America as kitchen utensils, they were replaced by electricity and became extinct by 1900.

At another level, dogs are more intimate members of the family, something clear in an anecdote, part of a collation of the continuity of mental life in animals and humans (Romanes, 1882). A contemporary of Charles Darwin, George Romanes (1848–1894) told of a dog that was used to accompanying a nursemaid and baby belonging to its mistress, on walks. On one occasion, the wind forced the nursemaid to draw her shawl over the baby and turn for home. However, her progress was halted by the dog becoming hostile, preventing the nursemaid from continuing, seemingly without the baby. The dog's faithful sentinel-like actions were only resolved when the nursemaid revealed the baby. Sometimes the relationship is very mutual and symbiotic: 'By degrees, the distance between Nira [a woman] and Majnoun [a black poodle] narrowed until each could anticipate what the other wanted. Nira could tell when exactly Majnoun wished to eat or go for a walk. Majnoun knew when it was time to comfort her, when it was time to sit quietly by her side' (Alexis, 2015).

At other times, we are unsure of the proper extent and limits of the relationship; friends and workmates can also be commodities, competitors, victims and pawns. Essentially identical animals can be pampered or persecuted, deliberately abused in violent domestic relationships, or the losers of our throwaway society. None of this is new. Since ancient times, dogs have had their behaviours altered by castration, teeth removed, or their vocal cords destroyed with hot fat. In some parts of the world, dogs are slaughtered, their meat a source of food and medicine and believed to bring good luck, health and vitality, its consumption part of a culture's identity (Podbersccek, 2009). Today's pets may be pampered but some suffer from separation anxiety, the distress evident in mournful whining, excessive vocalization, and destructive and abnormal behaviours when left alone at home (Schwartz, 2003; Sherman and Mills, 2008). And selective breeding for performance or looks has produced, either deliberately or accidentally, abnormalities such as breathing difficulties in bulldogs and hip dysplasia in sheepdogs. Finally, many unwanted dogs are surrendered, euthanized or abandoned, for diverse reasons: aggression and behavioural problems, hyperactivity, house soiling, fearfulness and escaping, being unfriendly to other pets and people, unsuitable or ill, or, commonly, being incompatible with their owners' circumstances, for example, housing restrictions, divorce, death, pregnancy, birth of a child, need to travel, allergies in the family, lack of time for an animal, finances or having been an unwanted gift (Scarlett et al., 1999; Protopopova and Gunter, 2017). 'Nevertheless, he is the one who holds the dog still as the needle finds the vein and the drug hits the heart and the legs buckle and the eyes dim' (Coetzee, 2000).

The reason for beginning this book about farm animal welfare with dogs is because dogs are our longest-serving animal companions. Dogs made us human, evident in some of our mythology. For example, to the Yarralin people of northern Australia, in the Dreaming, the dingo, or wild dog, and humans were one. Human ancestry, for them, is such that 'dingos are thought still to be very close to humans: they are what we would be if we were not what we are' (Rose, 2000). Others' myths also tell of humans born of the union of woman and dog-man, or of wolves that were once people. To the Black Tatars of Siberia, a naked dog was left to guard the first humans while their souls were being procured from heaven. In God's absence, the devil took the soulless people and defiled them with spittle, in exchange for giving the dog golden hair. God, on returning, turned the humans inside out so that we now have spittle and impurity in our intestines (Campbell, 1968). The dog is also an animal many of us are exposed to from our youngest days and throughout most of our lives. We continue to share our modern world with dogs in important, diverse and complex ways (Figure 1.2), the complexity of expectations of their place in society not unlike that of farm animals. It is a relationship we can easily understand, unlike that which many of us, in affluent western communities, have with today's farmed pigs, poultry, goats, cattle, sheep and deer. As we understand our relationship with the dog, it is hoped that we will come to understand our relationship with farm animals, a relationship as extraordinary as humans have with dogs. There is, however, one significant difference. The relationship we have with dogs is largely a

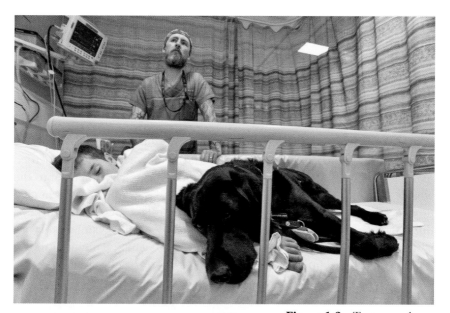

Figure 1.2 Two examples of how inseparable dogs' and humans' worlds can be. A young autistic boy in hospital with his ever-present assistance dog (*Photo courtesy Louise Goossens / Capital and Coast District Health Board*) and another youngster and his dog, surfing (*Photo courtesy Heather O'Brien*). (Available in colour in the plate section between pages 178–179.)

personal one, between individuals, whereas the relationship most of us, in the western world, have with farm animals is an impersonal but collective or societal one.

In evolving from a 'ferocious wolf' to a 'loyal companion', the dog epitomizes our evolution. We were once like wolves. Protohuman hominids, *Australopithecines*, who lived in Africa 6.5–1.5 million years ago, lived as vegetarians by gathering, or as omnivores, supplementing that diet by hunting small mammals, reptiles and insects. In the non-human sense of the word, *Australopithecines* were animals, primates without significant technical or cultural attributes (Finlayson, 2009). Finding ways to accumulate knowledge and power, practices and sciences, humans are smarter as a species than as individuals. And the extraordinary and synergistic relationship with animals, epitomized in our relationship with the dog, is part of our success, along with speech, morality and learning to live together, and farming, enabling us to develop markets and trade, become specialized, develop transportation systems and focus on efficiency.

Pigs, cows, hens, sheep, goats ... red deer and mankind

What began between wolves and humans eventually extended to other animals, and plants, as humans became farmers, herded animals, settled and ultimately transformed our physical and social environments into the worlds we now know. Just as humans and dogs were drawn together, so too were humans and sheep, goats, camels, reindeer, pigs, hens, turkeys, ducks, geese, guinea pigs, horses, donkeys, buffalos, zebus, llamas, alpacas, cattle, pigeons, water buffalos, guinea fowl, yaks, gaurs, banteng, rabbits, goldfish and carp among others, as well as more recently red deer. All domesticated – tamed and kept in a human-managed environment that controls their breeding, territory and food supply – a special relationship between animals and humans.

While not as old or as entwined as the relationship between man and dogs, the extent and diversity of the relationships are simply remarkable. Cows can be sacred or the source of milkshakes and hamburgers. Pigs hunt truffles for humans, are hunted by humans with dogs for sport and for food, or their carcasses are decorated as part of religious activities (the apple wedged between the jaws of a roast pig symbolized the rebirth of the sun in a pagan midwinter festival honouring Freya, a Norse goddess of fertility, Bezant, 1999). Chickens have been part of our sports and gambling worlds, are now our most significant source of protein (eggs and meat), and one of them, the rooster Chauntecleer, is the hero in the fable *The Book of the Dun Cow* (Humber, 1966; Lawler, 2015; Wangerin, 1978). Llamas, donkeys and camels have transported merchandise, horses have facilitated conquests of peoples and rats have been implicated in devastating plagues. The Romans farmed carp in ponds and the fish was popular in medieval times, while Fantail, Ranhu, Bubble eye and other variously shaped and coloured goldfish have become fashionable and popular ornaments (Balon, 2004; Teletchea and Fontaine, 2014).

One of the more remarkable illustrations of the nature of the relationship is the modern dairy industry. It was not only founded on the domestication of sheep and goats, and later cattle, but also on genetic mutations in humans enabling the digestion of lactose or milk sugar. Lactase, the enzyme that enables children and other young mammals to digest milk until weaning, now continues to be produced in older children and adults. The genetic variants allowing humans to continue to produce lactase, and thus digest milk, are now widely spread in different human populations (Cochran and Harpending, 2009).

Traditional farming, then, is a special relationship between species, an elaborated form of mutualism, a relationship where both parties benefit. The ancient or domestic contract (Kilgour, 1985; Rollin, 2008) – we take care of the animals and the animals take care of us – reflects the coevolution of both. Possibly having its origins in hunter-gatherer societies constraining the killing of animals to those needed for survival, and in a command not to kill for fun or self-aggrandizement, in return for being able to live peacefully with the spirit of wildlife (see Rodd, 1990), it is a hypothetical contract. While no individual animal makes a choice, the transition from the wild to the farm, with its many impacts and changes, is assumed to be reasonable. Richard Adams' (1920–2016) *Watership Down* (1972) in telling of the adventures of a small group of wild rabbits escaping their warren's destruction for urban development highlights the 'choice' animals have made. The rabbits came across another warren where the inhabitants were well fed and healthy, unnaturally big, strong and protected from elil, their enemies. Unlike the wild rabbits who lived as they pleased, these rabbits lived as the farmer, who managed them, wanted them to live. They happily ate carrots, kale, corn, lettuce, turnips or apples provided and benefited from not having to spend time foraging. Consequently, they developed a more sophisticated life compared with wild rabbits, even singing, dancing and becoming poets. These rabbits knew well enough what was happening but pretended all was well since the food was good, they were protected and had nothing to fear but being snared. Others, kept in hutches, lived dull and monotonous but safe lives. However, in becoming the 'elil's larder', the rabbits became strange, smelled of prosperity and forgot the ways of wild rabbits. The price they paid: a sadness and shame of their acceptance that some would be snared for their meat and skins. Domestication provided animals with shelter and protection from predators, and with better nutrition and health. In turn, humans had a more reliable and vastly increased source of food and other benefits. The symbiotic relationship leads to the belief that since humans are indebted to farm animals, humans have a duty to consider animals' interests and needs, which are essentially compromised for human benefit. Animals, unlike plants, are sentient and it matters to them what they feel and experience.

As land use and farming have changed from nomadic and settled extensive pastoralism to more intensive pastoral and even industrial-like systems, civilizations have progressed, and people have been drawn from an agrarian existence into urban ways of life. Our modern way of life has been contingent on the development of agriculture as much as civilization has helped shape farming. Agriculture fulfils several functions, among them the profitable, sustainable and safe production of food and fibre. However, that food has also to be safe and nutritious for whoever consumes it, and it must have been produced in a fair manner (Aiken, 1991). Modern farming not only has to care for animals but is increasingly having to also concern itself with complex issues such as how society makes use of, justifies and apportions its resources. These are choices that affect not only farm animals but also people, as illustrated in the following three examples of wildlife, coyotes, rabbits and badgers, enmeshed in the complex goals of farming.

Coyotes, rabbits and badgers

The crack and boom of a high-powered rifle echoing across the range reminded the escaping coyote of its place on a Colorado cattle ranch. Coyotes, animals that have adapted well to the modern North American environment, prey on calves and sheep, one of the risks of farming in the American west. Hot dry summers, cold dry winters, an annual rainfall of less than nine inches and snow during eight months of the year, Colorado's climatic and geographical attributes favour a thriving farming industry and it is one of the US's leading centres of beef production. The rangelands support grazing cattle and locally grown grain enables them to be finished on nearby feedlots. A typical ranch might run 450 breeding cows over up to 16,000 acres, some of the land leased from the government. Grazing is supplemented by hay from pastures irrigated from the snow-melt. While coyotes and the climate may be an imposition to raising cattle in Colorado, another threat facing the region's ranchers is the parcelling of federal or public lands for private use. Seeking a different lifestyle, many people have taken residence on 35-acre blocks, commuting to urban areas to earn an income. These blocks of land are used for little more than 'a house, a pond and a horse'. However, it is the number of these lifestyle blocks, evident in the long rows of roadside letter boxes in an otherwise secluded countryside (Figure 1.3), which also threatens the ranchers' livelihoods. Not only is there a loss of public land for grazing, but the associated increase in land values, and hence taxes or rates, makes it more difficult to farm profitably. The pattern of encroaching urbanization also affects the Colorado feedlot farmer. Not only lifestyle blocks raising land values and taxes but also their inhabitants, in objecting to the sights and smells associated with feedlot farming, help to make it difficult to continue to farm the way many have farmed for generations.

Half a world away, in the winter of 1997, some of NZ's farmers were torn between partying and being able to sleep more easily. The reason was

Figure 1.3 The long rows of letter boxes in rural Colorado (*Photo courtesy Ed Hansen*), dead rabbits displayed on a fence after the deliberate spread of a virus in rural NZ (*Photo: Otago Daily Times*), and a call to stop the cull of badgers in the UK (*Photo: Patricia Phillips/Alamy*), symbolic of the impacts of farming on people and wildlife.

the public revelation of the deliberate spread of a rabbit-killing pathogen. The virulent calicivirus causes rabbit haemorrhagic disease, decimates wild rabbit populations (Figure 1.3) and offered relief from rabbits competing for pasture with sheep and cattle. First reported in China in 1984 and accidentally released in Australia in 1995, the virus was introduced, apparently illegally, in response to the devastating environmental effects of an explosion in rabbit numbers. Invoking such terms as 'viral anarchy' and 'biological sabotage', its release was the inevitable response of normally law-abiding people to a belief in the rightful place of biological control agents, and to the frustration caused in placing the burden of responsibility for pest control on farmers in the face of falling economic returns (Parliamentary Commissioner for the Environment, 1998).

The badger has an iconic place in wildlife stories in the UK. *The Wind in the Willows* (1908), by Kenneth Grahame (1859–1932), is a classic children's tale of the adventures of Mole, Rat, Toad and Badger in, among other things, overcoming the nasty weasels. Symbolic of the countryside, badgers live in groups or clans in underground setts and are nocturnal, their diet as varied as earthworms, moles, raspberries and plums. However, it is via their role in harbouring and transmitting disease that the badger has gained most recent fame and notoriety. Once gassed to control the spread of rabies, badgers have been the subject of culling to manage the increasing risk of bovine tuberculosis in cattle herds, a disease also affecting humans. The effectiveness of culling, the availability of alternatives like vaccination against tuberculosis, and the impacts on badger welfare, have meant that the policy was and remains highly contested, with contrasting scientific, economic, ethical and emotional perspectives (Donnelly et al., 2006; McCulloch and Reiss, 2017). It is not surprising that there have been many public protests (Figure 1.3).

The sight of long rows of letter boxes, dead rabbits and protests against the culling of badgers, images from the US, NZ and the UK respectively, are symbolic of the complexity of issues affecting much of the affluent western world's most productive agriculture. Such images are also symbolic of the compromises we must make, including those concerning animals, for human benefit. We can, and indeed do object on behalf of the coyote, the rabbit and the badger, often without fully understanding why the animals are treated in that way. Although we may have compassion for coyotes, rabbits and badgers, to have compassion for farming requires acknowledging the impact of loss of farmland to lifestyle and urban dwellers, the economic necessity of controlling pests, and the risks that wildlife may harbour diseases affecting farm animals and humans. In the history of farming, they are but recent examples of balancing compromises to animals with benefits to humans.

Farmers today manage complex systems, making decisions being mindful of components as diverse as, for example, soil, water, plant, animal and environmental health and diversity, weeds and pests, the weather and climate, the advice of consultants and advisors, farm labour and contractors, customer requirements and satisfaction, their families, neighbours, community, and farmer groups or organizations, retirement, succession planning and the needs of future generations, regulations and government policies, as well as income, working expenses, profits, the economy and many other factors (Fairweather and Hunt, 2011). All these factors are drawn together in a network of interactions (Figure 1.4) in balancing benefits against consideration of the expectations for the treatment of animals. Similarly, the public may consider the important attributes of an ideal pig farm to include animal care, profitability, farm size, compliance with various rules and regulations, farm cleanliness, hormone, antibiotic and chemical use for increasing production, and workers' rights and welfare (Sato et al., 2017).

Increasingly, farming is being defined by its business focus, essentially ignoring or discounting its other aims since they do not contribute to the more quantitative aspects of productivity and efficiency (Rollin, 1995). In

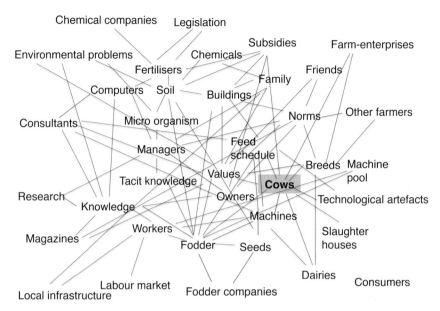

Figure 1.4 A simplistic portrayal of the internal and external relations making up a farm. Note that animals, in this case cows, are only one part of the network (from Noe and Alrøe, 2003).

order to share the wealth and supply the rising demand associated with urban prosperity, some in society have been willing to, or have had to, take risks in producing and distributing food (Body, 1991). Some risks have meant significant costs have had to be borne and there have been many eloquent portrayals of the animal welfare, human health and food safety, environmental and social consequences. Objections to the way animals are treated are not new. We learn of, and empathize with, the plight of the farm animals and farmers, and the fate of the coyote, the rabbit and the badger, in many ways. Not surprisingly, reflecting in part the complexity of systems of using animals, and in part our increasingly less sophisticated interactions with them, there are many objections to their fate, and the farming systems they are part of, especially regarding practices and systems not shared by the rest of the community.

There have been some elegant and reasoned critiques of farming. Narratives, especially fiction, by gaining our attention and arousing our emotions, help us to understand and interpret real-life social situations and provoke us to reflect on our morals (Boyd, 2009; Chambers, 1999). Stories invite us to see things from different perspectives and be creative in dealing with future events – they prepare us to adapt, to recognize and appreciate the intricacies and subtleness of relationships, enriching problems by placing them within the context of people's lives. For example, the plight of farm workers dispossessed of their land and livelihoods in the harsh economic times of the Great Depression, exacerbated by drought and high winds, and the relentless march of mechanized, industrial agriculture is told by John Steinbeck (1939) in *The Grapes of Wrath*. In more recent times, Ruth Ozeki's *My Year of Meat* (1998) is an exposé of the use of hormones in the beef industry, Jane Smiley's (1991) *A Thousand Acres* a narrative of changing family relationships and fortunes associated with 'corporate' agriculture and Michael Fox's *Superpigs and Wondercorn* (1992) an early commentary on modern biotechnology. Among the most influential have been Rachel Carson's (1962) *Silent Spring*, detailing the persistent environmental effects of pesticides such as DDT and Ruth Harrison's (1964) *Animal Machines*, which exposed the cruelty of intensive methods of raising pigs, calves and poultry. Before them all, Anna Sewell's 1877 tale of *Black Beauty* told of the hardships faced by London-taxicab horses, while more recently, James Rebanks' *The Shepherd's Life* (2015) is an account of farming, shepherding and living in England's Lake District. Science fiction, for example, Philip K. Dick's (1968) *Do Androids Dream of Electric Sheep?* and H. Beam Piper's (1962) *Little Fuzzy*, explores the relationship between man and animals in a manner less threatening than in everyday settings (Clements, 2015), the former by contrasting humans and androids, the latter a race of people on a distant planet colonized by humans.

Similarly, there have also been critiques, many detailed and sophisticated, of animal welfare and Table 1.1 contains a selection.

Animals like dogs, pigs, calves, poultry, badgers, rabbits and coyotes experience varying degrees of thwarted preferences, discomfort, suffering

Table 1.1 *A selection of works addressing farm animal welfare.*

Work and author
Animal Machines. Harrison (1964)
Animal Liberation. Singer (1975)
Agricide. The Hidden Farm and Food Crisis That Affects Us All. Fox (1986)
Factory Farming. Johnson (1991)
Animal Welfare. A Cool Eye towards Eden. Webster (1994)
Farm Animal Welfare. Rollin (1995)
Animal Welfare. Spedding (2000)
Dominion. The Power of Man, the Suffering of Animals, and the Call to Mercy. Scully (2002)
Eternal Treblinka. Our Treatment of Animals and the Holocaust. Patterson (2002)
Animals Make Us Human. Creating the Best Life for Animals. Grandin and Johnson (2010)
Farmageddon. The True Cost of Cheap Meat. Lymbery and Oakeshott (2014)

and death, compromises that inevitably bring a variety of benefits for humans. In interacting with animals, we cannot escape the fact that for our lives to go well, animals' lives are sometimes enhanced, often compromised.

Animal suffering

Animals can be affected in a variety of ways (Fraser and MacRae, 2011). Humans, both knowingly and unknowingly, change animals' habitats. For example, ploughing land and harvesting crops can suffocate animals like rodents in their burrows or make them more susceptible to predation by removing cover. The introduction of exotic species and diseases, exposure to pollutants and toxic chemicals, soil and water degradation, deforestation and habitat loss, all change ecological systems providing opportunities for some animals, for example, coyotes and rabbits, and death and extinction for others. Finally, human activities and structural changes like urban development, vehicles and roads, towers, windows and artificial lighting may disrupt and harm many animals. Such indirect harms are rarely given consideration in animal welfare unless they impact on iconic wildlife conservation. In addition, we directly influence animals by farming them, as well as keeping them as companions, and so forth. Often this involves deliberately harming them, such as exposing them to husbandry procedures, as well as slaughtering them. Animals not only suffer from physical pain and distress, as frequently articulated, but pain also includes a range of unpleasant states, such as fear, thirst and hunger, frustration, aggression, boredom, loneliness, anxiety, nausea, itching, breathlessness, exhaustion, temperature extremes, grief, and so forth. Both physical and emotional forms of suffering are important.

The causes of suffering inherently associated with farming include distortion of social and kin structure by culling and grouping, limiting or preventing movement and migration, less varied food and nutrients, curtailment of parental care of young by separation and weaning, or exposing animals to circumstances they find adverse or would not choose for themselves. Many such harms are also inherent in wild populations of animals. For example, the slow death at the hands of a predator, or the lack of sufficient resources such as food and shelter, over winter. Procedures to increase or intensify animal production and efficiency can not only enhance, but also compromise, animal well-being.

Animals can thus suffer for a diversity of reasons. At the most superficial level, if they are unable to find, prevented from obtaining or not supplied with the resources they need in the environment in which they find themselves, then they are likely to be at greater risk of suffering. These needs, covering all the factors likely to influence the welfare of farm animals, have been expressed as ideal states (Webster, 2005a) whereby animals should have:

- freedom from thirst, hunger and malnutrition;
- freedom from discomfort;
- freedom from pain, injury and disease;
- the freedom to express normal behaviour; and
- freedom from fear and distress.

Neglecting or failing to provide for the physical, health and behavioural needs causes pain or distress. Although what the animal experiences is similar whatever the human motivation or justification for its treatment, it is convenient to distinguish between unnecessary or unreasonable and necessary or reasonable suffering. In other words, animals can be treated differently for a variety of reasons, from necessity and convenience, to thoughtlessness and ignorance, and indifference and sadism.

Unnecessary suffering is sometimes the result of people being unaware, ignorant or indifferent to animals' needs, or even entertained by their suffering. Mark Twain's *Adventures of Huckleberry Finn* (1884) tells of loafers, idle people who avoid work, who could not be woken or made 'happy all over' unless being entertained by a dog fight or by 'putting turpentine on a stray dog and setting fire to him, or tying a tin pan to his tail and see him run himself to death'. In more recent times, videos of the athletic responses of cats, startled by the placement of a cucumber on the floor behind them, gained huge interest on social media (Animalia, 2016). Part of that interest included consideration of whether the cats suffer unnecessarily or whether they were being provided with a more enriched environment.

Animals may also suffer if people in charge of them are unable to care for them because they are experiencing their own problems. There is a clear association between farmers experiencing physical and psychological difficulties and being unable to care for their livestock. Ill-health, including psychiatric problems, difficult financial circumstances or personal and family relationships, adverse climatic conditions (e.g. droughts, blizzards or floods) or inadequate family or social pastoral support, have all been implicated. There are some insightful lessons from several reports. For example, a NZ animal welfare inspector reflecting on over a decade of investigating such events as: '500 dead sheep, all of which died for the want of food, or shearing or both. It was all the owner could do to pick up the readily accessible dead sheep then retire back to bed … The signs of this man's ill-health were obvious …' The common factors with this and other cases were that the owners were male, they had not sought help, all were sick and depressed, were under intolerable pressure and had ceased to function rationally. Interestingly, perhaps disturbingly, there was a degree of public knowledge of what was happening, but no one intervened (Presland, 2011).

Similarly, interviews with farmers in Ireland (Devitt et al., 2014), who had been investigated about animal welfare incidents, revealed they had had difficulties with their advanced age and in getting help on the farm. 'I couldn't get the animal out [of the water] and there was no-one around to help, and the animal just disappeared into the water ... It's how things are these days; my son has emigrated, my neighbours also are elderly.' The farmers had mental health-related problems and stress was prevalent, impacting on their ability to farm. In addition, they had differing perceptions of animal welfare, and poor uptake of support services because they were resistant, self-reliant and unable to talk about mental health. The farmers themselves commented that farming was difficult and can be very stressful: '... I borrow more money and you try again, but I can't take too many more bad years'; '... four of my [agricultural college] classmates have done suicide – four people that I know.'

While there was a low risk of animal neglect, the risk factors, at least in industrial farming in Denmark (Andrade and Anneberg, 2014), were financial, divorce and psychiatric problems, technological breakdown, family problems and concern with growing governmental control. One farmer's reflection was:

> ...farmers were making money. I borrowed money, drew up a budget, it all looked good if the price didn't fall ... But then when I was going to sell the first lot of pigs, it was all upside down. Prices were falling, the expenses for fodder went up ... The old stalls started to break down. The pigs broke everything ... Normally a farm with 8500 slaughter pigs should be managed in 3–4 hours per day, but it took me 10–12 hours, and I was constantly behind.

Some of the reasons given included concern at the growing expectations for control of farm animal production, for example, through legislation, inspections and audits. The authorities, veterinarians and the public now have different, even opposing views of acceptable practices, views that differ from the practices they, the farmers, and previous generations, regarded as good, further isolating and challenging those individuals.

And in the US, there is concern that while some individuals may cut corners, others may be pushed into it by the circumstances in which they farm, their production systems and supporting infrastructure, and retailers, for example, vertically integrated intensive systems, which constrain choices and increasingly make it difficult for farmers to make their own decisions (Hendrickson and James, 2005).

The primary topic of concern expressed in calls to the UK's Farming Community Network Helpline was, by far, financial issues, also raised in combination with animal welfare (Farm Animal Welfare Committee, 2016). Underlying factors also included:

- advancing age reducing an individual's ability to carry out everyday activities;
- limited availability of farm help sometimes attributed to a decline in rural, small-scale farming systems;
- overstocking;
- experiencing difficulties with technology;
- having different perceptions of acceptable standards of welfare;
- the cost of addressing animal ill-health resulting in greater tolerance of poor animal welfare;
- more complex issues related to mental health and depression (indecisiveness, compassion fatigue and an inability to recognize and deal with their circumstances); and
- fear or feelings of failure and a lack of knowledge or support.

Human stress was a prevalent theme, with a clear link between instances of physical and mental ill-health and the ability of some farmers to care for their animals as well as themselves. Not surprisingly, high rates of occupational injuries and suicide are evident in many countries (Donham and Thu, 1993; Fraser et al., 2005; Gallagher et al., 2007). The barriers to individuals not obtaining support, both medical and agricultural, include resistance, self-reliance, stoicism and an aversion or inability to talk openly about mental health. One of the values of being a good farmer is in being strong and self-reliant, able to deal with problems. Individuals are not supposed to succumb to pressure, typically not admitting problems, even to their families.

Hiding their frustration from others was also evident in at least some cases of cruelty towards dairy cows in NZ. Farm employees, unable to control their anger and frustration with the stress of long hours, family pressures, financial burdens and fatigue, redirected their hostility by twisting and breaking tails (e.g. MacLean, 2016). It is a reminder that abusive treatment of animals is sometimes the consequence of our relationship with other humans. Those convicted were otherwise of good character and standing in the community, and generally cooperated and accepted responsibility. Other motivations for animal cruelty may also arise, or be redirected, from more violent relationships with other humans. For example, abuse of, and threats against, companion animals are part of domestic violence, used by males, usually, to intimidate females and children. Similarly, abused children may abuse animals to satisfy a need to control and dominate others, and adolescent males torture animals in response to social rejection or to gain peer approval. Empathy towards animals is likely to be distorted by such cruelty (Flynn, 2001). Although commonly involving pet dogs and cats, farm animals, cows, sheep, chickens and goats have also been threatened, injured or killed (Ascione et al., 2007; Roguski, 2012).

Cruelty, then, is socially unacceptable behaviour that intentionally causes unnecessary pain and suffering of an animal, the domain of social deviants, or the 'bad, mad or sad'. Not surprisingly, society's response is to prohibit these actions and there are considerable efforts undertaken to expose, educate, help or prosecute. Laws, regulations, codes of welfare, guidelines and assurance programmes all articulate society's expectations of how animals should be treated. However, some unnecessary or unreasonable animal suffering or neglect is clearly multi-causal with work and family pressure, financial and personal problems leading to stress making routine farm work difficult and, in turn, leading to animal neglect. Inadequate technology, family and community support, and animal welfare inspections also contribute along with changing social expectations, especially those dictated by others. Instead of just imposing stricter regulations, requirements for inspections, and demands for evidence of competence or professional training, such actions are arguably irresponsible without some acknowledgement of the underlying contributing factors. Perhaps there is scope for understanding and changing the attitudes both within farming and beyond so that demands for farm animal welfare are fair and equitable for both animals and people.

In contrast to unnecessary suffering, necessary or reasonable suffering constitutes 'the overwhelming majority of animal suffering at human hands' (Rollin, 2006). It is not the result of neglect or cruelty, but is suffering associated with normal or acceptable animal use. Necessary suffering is the result of exposing animals to conditions or practices that society agrees are acceptable for the benefit of the animal, farm management, humans or ecosystems. For example, some sheep are routinely tail docked when young for the overall benefit of the animal and/or farm management to reduce the risk of faecal soiling and flystrike and make wool removal easier. Similarly, males of several farm species are castrated to modify their behaviour making them easier to manage, or to prevent meat being tainted by male hormones produced after puberty, and cattle have their horns removed to reduce the risk of injury to other animals and humans. All these procedures produce acute pain with pain relief rarely provided, principally for pragmatic or economic reasons. While the harms and benefits of these husbandry procedures are well documented, there are many other compromises resulting in animals experiencing suffering – the work that they do, the social disruptions they experience, the confinement and boredom they are exposed to and the resources that they are deprived of. These too are all deemed necessary and reasonable because of the benefits they bring to humans. For example:

- high-producing dairy cows may experience feelings of hunger, tiredness, fullness and sickness (Webster, 2005a) and/or be separated from their young at birth for more efficient milk production;
- segregated early weaning of piglets, to reduce the impact of growth-depressing pathogens, resulting in unusual behaviours such as nosing the bellies of other piglets, increased activity and aggression, and increased risks of intestinal disorders (Fraser, 1978; Robert et al., 1999);
- the placement of fences to best manage pasture may prevent animals from access to preferred resources like high ground, shelter or isolation;
- handling-induced fear associated with treatments (e.g. vaccines, insecticides) to prevent or alleviate ill-health;
- deprivation of feed and water to facilitate transport and slaughter.

These and other compromises, the 'freedoms' animals give up, are considered justified for the benefits that accrue to humans – those managing animals, those involved in moving and processing them, and those either directly or indirectly involved in or using and consuming the products. This form of suffering, justified by the benefits accruing, is generally accepted in that society agrees that the benefits to the animals, farm management, food processing and consumers or society outweigh the harms to animals.

Ruth Harrison (1964) once noted that the difference between necessary and unnecessary suffering was related to the number of people who benefited:

> ... if one person is unkind to an animal it is considered to be cruelty, but where a lot of people are unkind to a lot of animals, especially in the name of commerce, the cruelty is condoned and, once large sums of money are at stake, will be defended to the last by otherwise intelligent people.

The feelings animals experience from necessary and reasonable compromises may be no different to those experienced by animals being neglected or abused. It is just that the human motivation differs. Neglect and abuse, or unreasonable and unnecessary practices, are socially unacceptable and open to punishment, whereas compromises justified for human benefit are taken as reasonable and necessary and socially sanctioned. The difference between reasonable and necessary, and unreasonable and unnecessary, the motivation for treating animals in certain circumstances, is part of this book. As humans determine how farm animals are treated, understanding what drives human behaviour may help identify ways and means of managing our impact on animals. In other words, there are a variety of overlapping drivers of animal suffering. Knowing and understanding them, the complex part of animal welfare, is critical to addressing their impact, and to the future of animal welfare.

Animal suffering is not just the realm of rascals, social misfits and rogue farmers but sometimes, perhaps usually, also a reflection of the system itself. The benefits of farming are inextricably linked to people, the market and economic growth impacting on animal welfare. The complexity of underlying reasons contributing to animal suffering also means we cannot use the same response, for example, educating, exposing or prosecuting people. Cries of 'farmers need to be educated' or 'if farmers can't look after their animals then they shouldn't be in the business' may sometimes be valid, but without a deeper understanding of the context or circumstances of less than optimal animal performance and welfare, society is unlikely to realize sustainable and equitable solutions for both farmers and their animals. However, it is the animal's point of view that is important, so how do we understand that?

What is animal welfare?

At its simplest, animal welfare is about what the individual animal experiences, how it fares, since to fare well is to have a good life (Green and Mellor, 2011). Animals are sentient and experience pleasure and suffering, and a range of feelings (joy, frustration, fear, hunger, playfulness etc.) that matter to them. These feelings are, in turn, dependent on the animal's health, the absence of disease, injury, suffering and distress, and the provision of resources such as food, water, shelter, comfort, opportunities and interactions with others. When unable to obtain, or prevented from obtaining, those resources, the animal suffers, especially if the insult is severe, prolonged or complex. The simplest definitions of animal welfare, then, are that the animal is 'fit and feeling good' (Webster, 2005b), is 'functioning well and feeling well' (Nordenfelt, 2006) or is 'healthy and has what it wants' (Dawkins, 2008).

Table 1.2 *A selection of terms commonly used to reflect expectations of how animals should be treated (Fisher et al., 2014).*

Aesthetics	Beauty or the appreciation of good taste
Dignity	To be worthy of esteem or respect
Integrity	The state of being whole or undivided
Intrinsic or inherent value	The value an entity possesses as an end in itself, regardless of its utility
Respect	To admire someone or something deeply, as a result of their abilities, qualities or achievements; to have regard for the feelings, wishes or rights of others
Reverence	Deep respect for someone or something
Rights	A moral or legal entitlement to have or do something
Telos	An ultimate object or aim, the nature of something

The different human understandings of animals' needs, and of the value of the benefits humans seek from animals, are reflected in different understandings of animal welfare (see Fisher, 2009). Common perspectives include what the animal experiences or feels, how it performs (how fit it is), or whether it is living a reasonably natural life carrying out natural behaviour in a natural environment and able to satisfy the biological and psychological needs and interests that matter to it. A final perspective relates to the animal being treated with dignity and respect and there are several terms reflecting expectations of how animals should be treated (Table 1.2).

Perhaps some of these terms are more symbolic than practical, of limited value in providing a foundation for practices and laws, but they nevertheless capture something important. Concern for an animal's integrity or dignity is not routinely captured by reference to animal health and welfare, a topic more concerned with what the animal experiences. Consider that wild and farmed animals may equally experience poor welfare, but concern for an animal's dignity and integrity demands that humans respect farm animals, as well as treat them humanely (Christiansen and Sandøe, 2000). This insight may assist the apparent inconsistency in dealing with terms such as natural. For example, modern farm animals such as the dairy cow may not be 'natural' in producing large volumes of milk, having her calf removed shortly after birth, being housed, fed with manufactured rations and provided with unnaturally high planes of nutrition during winter. However, the modern dairy cow still has many features of an 'ancestral' or 'natural' cow, suggesting that it might be motivated to perform activities regarded as natural (e.g. grazing, resting and sleeping, and investigative, social, sexual and maternal behaviour). The idea of animal welfare reflecting natural behaviour emphasizes that, in addition to minimizing harms (e.g. pain and distress, frustration, aggression), animals also need, or at least should have the opportunity, to experience pleasurable or positive behaviours, as they would under natural conditions (Brache and Hopster, 2006). While we all know generally what animal welfare means, it is worth noting that we may not all share the same nuances, for example, there are reputedly 66 different meanings of the term 'nature' (Preece, 1999) – presumably 'natural' is similarly diverse. However, if societal views, for instance those held by consumers (Autio et al., 2018), reflect these more abstract concepts then, it is argued, those involved in animal welfare, be it ethics, husbandry or science, should too (Röcklinsberg et al., 2014). The different perspectives or understandings are also apparent in the definition of the term 'animal welfare'. It was first used in the 1960s as 'a wide term that embraces both the physical and mental well-being of the animal ... [that] must take into account ... the feelings of animals that can be derived from their structure and functions and also from their behaviour' (Brambell, 1965). With the rise of animal welfare science, not surprisingly, there have been many attempts to define the term (Table 1.3).

Definitions have various applications, ranging from helping understand and provide guidance, to setting boundaries, and differentiating and

Table 1.3 *Examples of the range of different understandings of animal welfare (Fisher, 2009).*

Definition or understanding of animal welfare
The absence of suffering
Fit and happy or feeling good
A state of being in which at least the animal's basic needs are met and suffering is minimized
Animals are healthy and they have what they want
The state of an animal as it attempts to cope with its environment
A broad predictive physiological and behavioural capacity to anticipate environmental challenges
A state of complete mental and physical health, where the animal is in harmony with its environment
A state which includes some measure of a successful life
The use of methods for handling and management that impose the least amount of stress or distress
The viewpoint that it is morally acceptable for humans to use non-human animals provided unnecessary suffering is avoided
Ultimately an economic or socio-political issue, primarily a subjective matter of human perceptions, a subset of human welfare since people's preferences determine their actions

excluding unwanted concepts. They also highlight different viewpoints, insights and understandings. Conversely, the limited understanding that a single, precise definition implies would not seem to address many of the issues of concern or deal adequately with the multidimensional nature of animal welfare (Rushen, 2003), risking excluding common understanding and the erosion of the moral element of the term (Stafleu et al., 1996; Tsovel, 2006). Thus, different definitions bring advantages and disadvantages to understanding an animal's state and how it should be treated. Given the value of different perspectives in understanding complex issues, there is merit in considering acceptable animal welfare as a judgement based on as many different sources of information and understanding as possible. For example, the goals of agriculture, the perspectives of all interested parties, the skills of good stock handlers, the animals themselves, and our future demands on and expectations of animals (Fisher and Mellor, 2008). We make sense of, and understand, the world in many ways. For example, through biophysical and social science, case histories or narratives, and through feeling, or empathizing with and understanding different and changing views. Perhaps a consistent definition does not matter if we all know essentially what animal welfare means and are prepared to take different understandings into account. In any consideration, it would be pertinent to at least acknowledge these different understandings and their influence on expectations of how animals are treated. This is especially so if human-animal values are increasingly being shaped by interactions with pets, arguably members of the family, while interactions with 'real' animals are increasingly absent from modern life. While there are different definitions, focussing on what keeps animals healthy and what they themselves want may enable animal welfare science to focus on its strengths – evidence and the testing of hypotheses (Dawkins, 2016). However, while feelings are still integral to much of our understanding of animal welfare, they are difficult to describe and measure, and we resort to indicators of what the animal is experiencing.

Mammals have a range of physiological and behavioural mechanisms that allow them to adapt to different situations, keeping the animal within a narrow range of environmental limits most conducive to survival. Some of these are quite ingenious. For example, Cape ground squirrels, at home in the hot, dry and shade-less regions of southern Africa, raise their large, flat, bushy tails for shade and turn their backs to the sun whenever the temperature exceeds 40°C. Such behaviour means they can forage for up to 7 hours a day compared with only 3 hours without shade (Bennett et al., 1984). While farm animals rarely encounter such challenging environments, they do have a range of ways of reducing heat load and maintaining their core body temperatures within the limits necessary for normal

The simplicity and complexity of animal welfare • 23

Figure 1.5 Beef cattle sheltering from the sun under trees, sheep crowding together to use each other's shade and that provided by fences, and an example of the importance some people place on providing shade to farm animals (*Photos: Mark Fisher*).

functioning. These include physiological responses like vasodilation to dissipate heat from the blood, reducing heat production, and physical responses like panting, sweating, reducing activity and seeking shelter (Figure 1.5). For instance, sheep seek shade, sweat more, pant to lose heat, drink more and eat less.

At the other extreme, farm animals must also adapt to, or contend with, cold conditions. Snow, rain and wind can make life miserable and animals also have an array of physiological and behavioural responses to deal with cold, wet conditions. They include diverting blood away from the extremities to prevent heat loss, postural changes like hunching up, and seeking shelter. Heat production is increased by shivering and raising the basal metabolic rate. If these responses are required to a significant degree, resources may be diverted from growth and production and health and welfare may be compromised. These are all homeostatic mechanisms enabling the animal to cope with its environment.

Perturbations or changes in the environment to which farm animals are exposed include, in addition to heat and cold, feed and water deprivation, injury and disease, and a myriad of human interventions from the proximity of people to handling, transport and the undertaking of painful husbandry procedures (mutilations) that can result in fear, pain and distress and the lack of opportunities to behave normally. The measures that animals use to adapt to, habituate or respond to these challenges are generally behavioural and/or physiological. For example, when newborn red deer calves are in danger of being disturbed they typically freeze or become immobile, with their necks extended along the ground, their ears flat, and remain silent. Their heart rate falls dramatically (Espmark and Langvatn, 1979, 1985). This adaptive hiding behaviour has evolved as part of an antipredator strategy to remain inconspicuous during the first few weeks of life.

Heart rate and shade or shelter seeking behaviour are examples of homeostatic mechanisms that animals, and humans, use to cope with less than ideal as well as adverse environments. As well as heart rate, there are a host of physiological measures used to infer what an animal is experiencing, either pleasure or pain. Similarly, there are many behavioural insights from observing animals' responses in natural and novel situations (e.g. feeding, nesting, grooming), to asking them to work for resources to understand how important those resources are. There is no single measure of welfare and different measures may give different insights. Some measures are reasonably objective, for example, live weight; others are based on more subjective observations of animals' body language, for example, describing animals as content, happy, distressed and so forth, and subjecting the data to analysis revealing positive to negative welfare and active to passive scales (Wemelsfelder et al., 2001). Consequently, a range of scientific measures are best used to understand if the animal is fit and healthy, has what it wants and is feeling good. Some measures are novel and innovative, for example, monitoring tear staining in pigs as they 'cry' more when injured or reared in a poor environment (DeBoer et al., 2015; Telkänranta et al., 2016). Some challenges can be innocuous and relatively easy to deal with, and some tolerable though requiring greater effort. However, some are aversive requiring major changes that can compromise fitness or, if extreme, survival. Changes in these homeostatic mechanisms are thus frequently used as indicators of animal welfare (Table 1.4).

Some of these adaptive changes or homeostatic mechanisms are well described but the impact on what the animal feels is less well known. For example, sheep can be deprived of feed prior to slaughter to limit the accumulation of effluent during transport, reduce the risk of microbial contamination during slaughter and to facilitate more hygienic evisceration. Fasting results in physiological changes as energy derived from the diet is replaced by energy mobilized from body fat and muscle. Changes in circulating metabolites – for example, free fatty acids and ß-hydroxy butyrate – have been used to infer hunger, but the extent of that hunger, how the animal feels, is not well understood (Fisher et al., 2015).

Table 1.4 *Examples of homeostatic mechanisms used as indicators of animal welfare and an example of how they relate to thermal stress (from Broom and Johnson, 1993; Fisher, 2007).*

Homeostatic mechanisms used by animals to attempt to cope with environmental changes

Challenges	Pain, anxiety, handling, novel situations, proximity of people or predators, transport, social isolation, crowding, thermal stress, aggression
Behavioural responses	Orientation, startle and reflexes, flight, struggle
Physiological responses	Heart rate, blood pressure, respiration rate, body temperature, hormonal and metabolic changes, appetite
Long-term responses	Changes in live weight, reproductive success, longevity, susceptibility to disease and injury, inappropriate or pathological behaviour (e.g. stereotypies)

The responses of animals to different stages of thermal challenge

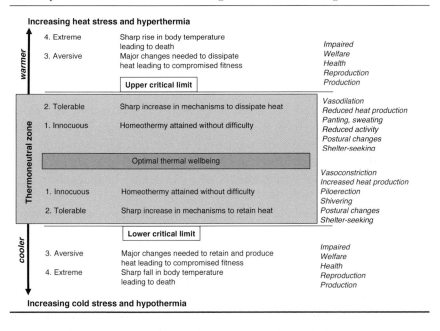

Whatever the difficulties and possibilities for understanding animal welfare, it is important to remember that animals normally adapt to the circumstances they encounter. It is likely that the ability to adapt to the changing environment at the end of the last ice age favoured those animals that would then become domesticated (Coppinger and Smith, 1983).

However, where those changes are excessive, prolonged and maladaptive and the animal is unable to do something about its circumstances, either behaviourally or physiologically, contributing to pathological harm, the animal fails to or has difficulty adapting and suffers. While fit and feeling well is a good description of how well an animal might be doing, feelings are difficult to measure, be they animal or human. Although we can, with some degree of probability, think we know what others, either animals or humans, feel, strictly speaking absolute understanding is difficult if not impossible. This is especially so of animals more evolutionary and ecologically removed from say the dog, from foxes to fruit bats and fish. Thus, physiological and behavioural indicators of welfare (Table 1.4) are used to infer how well the animal is doing, placing science in a powerful position to help determine what are acceptable and unacceptable practices involving animals. It is well to remember that many of these indicators can mean the animal is adapting or coping with the circumstances or environment it is exposed to, and not that it is necessarily suffering. Animal welfare then is determined by an animal's ability to adapt or cope without suffering, or to have a good life. Unfortunately, changes in homeostatic mechanisms, designed to enable the animal to cope, are sometimes uncritically used as a proxy or indicator of reduced animal welfare. This is not necessarily the case; perhaps only when the animal fails to cope can animal welfare be said to have been compromised. And failing to cope encompasses states as different as comfort, production and survival.

While animal welfare is, then, a simple concept, that is, how an animal feels is often indicated by how it performs and behaves, it can be remarkably complex when you determine how it feels and how we understand, balance and value the benefits to humans against the costs to animals. Animal welfare moves into the social domain when its welfare is compromised for human benefit and must be justified. The term encompasses aspects of natural selection, coping, well-being, satisfaction of preferences, fulfilment of needs, and natural behaviour, or a combination of them (Nordenfelt, 2006), with what suffering we determine to be reasonable and necessary that, in turn, depends on our beliefs and prejudices. In other words, it is not just about the animal but also about our understanding of the animal, what it is experiencing and what matters to it, as well as the justification for any pain and distress caused. It is foremost a social construct reflecting societal values (Fraser, 1995, 1999; Rushen, 2003; Sandøe and Simonsen, 1992; Tannenbaum, 1991) and thus a complex, multifaceted public policy issue that includes important scientific, ethical, economic and political dimensions (Petrini and Wilson, 2005). Animal welfare, then, has an animal dimension and a social or people dimension. To discuss, write or decide about either without being aware of and acknowledging or incorporating the other is poor. To address both is to acknowledge the complexity and reduce the confusion and contention in animal welfare.

'Pigs are easy, people are the challenge'

The different understandings and definitions of animal welfare have at their centre the good of the animal. They differ in how that good is understood, described or assessed, and valued – often, and increasingly, but not exclusively, by scientific means – and/or how compromises to the good are deemed necessary and reasonable or justified, by appeal to values. Animal welfare then is not just a technical concept or the state of the animal, but also a reflection of, or shaped by, underlying human values. We cannot confirm or refute how animals should be treated using empirical knowledge alone. In other words, animal welfare is about people as much as it is about animals. It is tied to the cultural practices we share with others in our community, our expectations of animals be they sources of companionship, food, transport, sport or whatever. Our views are also influenced by our beliefs and prejudices, about both what is important to animals and what is important to us. As we will see in the next section, there are many examples of apparent inconsistencies in understanding and treating animals borne out of our differences in our values, beliefs and prejudices. There are also many examples of how we use our common sense and values to understand animals. For example, the hormone cortisol is typically secreted in farm animals in response to pain, such as that accompanying castration. However, it is also secreted, often in a remarkably similar way, in response to mating. Transport is another stressor known to elevate circulating cortisol concentrations – in sheep being trucked to slaughter it may well represent discomfort or suffering, in a dog being taken for a family outing it may reflect excitement and pleasure.

The title of this section, taken from a contribution to the effects of agricultural intensification on rural communities (Thu and Durrenberger, 1998), highlights the fact that animal welfare is about people as much as it is about animals. Social expectations, the influence of people on an animal's welfare or at least the expectation of how it should live its life, can be both simple and complex because we are all different. Consider your responses to the following examples. First, three images of lambs being suckled (Figure 1.6). It is suggested that each of the lambs is, overall, faring well. However, the feelings they evoke can be quite different. When published, without a caption, in the *Utne Reader* in 1988, the woman suckling the lamb drew responses ranging from repulsion at the bestiality and pornography displayed, to compassion, the beauty of being in

Figure 1.6 Three different images of lambs being suckled, each arguably faring well, yet each image evokes quite different feelings or emotions in humans. A ewe suckling its lamb (*Photo: Mark Fisher*), a cow suckling two orphan lambs after they bonded when their mother was unable to care for them (*Photo: Stephen Barker/Rex/Shutterstock*) and Catalin Valentine's Lamb, Ancash, Peru (*Photo: Rosalind Solomon, courtesy Harvard Art Museums/Fogg Museum*).

touch with nature, and a tender antidote to the cruelty of factory farming (Anonymous, 1988).

A second example, and perhaps one of the more curious examples of human understandings of animals, is the capybara. The world's largest living rodent weighs around 50 kg, stands about 60 cm at the shoulder and has dark brown to reddish fur. Capybara are relatively docile animals, native to the tropical and subtropical areas of South America. A grazing herbivore, it inhabits a variety of lowland environments containing ponds, lakes, rivers, streams, reservoirs, swamps, savannas and wetlands. Capybara are hunted as well as ranched or farmed in much of the region and their meat is a subsistence resource for indigenous poor and minority ethnic groups, as well as being sold to wealthy consumers. The animal's leather is used in footwear, clothing, luggage and handicrafts, mainly in Europe. However, the capybara's curious relationship with humans is because of its nocturnal and aquatic habits – it was classified as a fish by the early European explorers to South America. Consequently, and sanctioned by religious decree, large quantities of capybara flesh are eaten by urban Roman Catholics during Lent in Venezuela (Moreira et al., 2013; Nogueira and Nogueira-Filho, 2012). Interestingly, this is not an isolated example – birds that live in aquatic environments and eat mainly fish have been regarded as fish, for example puffins in medieval England (Vialles, 1994). And, in the Middle Ages, monks apparently ate newborn or rabbit foetuses (laurices), also during Lent, as they were considered an 'aquatic dish' (Lebas et al., 1997).

That animals evoke different responses inevitably means people treat them differently – what we believe determines how we act (and how we act helps determine what we believe). What these examples do is remind us that the place or treatment of animals is laden with values. Not only does animal welfare consist of what the animal feels or experiences, inevitably reflecting its biology or its biological needs, or as Rollin (1990) notes, it is common sense that fish swim and birds fly, as well as our expectations of animals as sources of companionship, food, transport or whatever. These examples also expose our innate and culturally ingrained prejudices about both the needs of animals and those animals' roles and value to us, be they economic, political, spiritual or whatever.

Thus, animal welfare is not just something that can be measured objectively. Although measures can be used to inform our understanding and decision-making, relying exclusively on evidence both obscures the subjective and value-based issues involved, and helps to give a deceptive impression of the precision of those measures. Humans are good at making subjective judgements. We know when our children are happy, when our pets are content. For the most part, we also know when animals are being harmed or are content. Clearly, we can also use evidence to understand what animals are feeling and help justify whether their lives are going well. This, then, is the complex part of animal welfare, what we think about animals, what their value is to us, what our culturally ingrained prejudices are towards them and how we justify their needs against ours. In short, how we understand the complexity of the various relationships between animals and mankind. And it is this complexity, or at least the failure to acknowledge it in its fullness, that is fertile ground for misunderstanding. Failing to acknowledge mankind's dependency on the natural world, our love of life or biophilia, and consequently the risk of becoming dissociated from, even contemptuous for, some relationships. It is not just ignorance; the complexity is real, the problems are wicked. The challenge that inspired the title of this section was that, 'if there is a people problem, it is the people who work from the assumption of the business schools rather than the realities of farmers'. Because our worlds are complex and our roles relatively specialized, few of us understand the farming any more. But we are exposed to accounts, opinions and reflections that reflect pragmatic, emotional, professional or whatever bias. They, along with the cultural, economic, spiritual, religious and other dimensions, help to make animal welfare a complex and 'wicked problem'.

Animal welfare is wicked

Difficult to describe, complex, changing and subject to inconsistencies and considerable political debate, wicked problems are intractable and elusive because they are influenced by many dynamic social and political factors as well as biophysical complexities (Rittel and Webber, 1973; Whyte and Thompson, 2012; Peterson, 2013). They are not easily solved as the capacities to address them are widely distributed, they are difficult to predict and they involve conflicting goals (Jones, 2011). They are at best managed and progressed through understanding, shared meaning (Bohm, 1996) and compassion. They require knowing the animal, for example, the coyote in Colorado, knowing the circumstances of the compromises to its life and those of the farmers competing with it, and knowing the long rows of letter boxes are symbolic of the factors affecting local farming and thus the welfare of farm animals. The immediate face of the problem, the coyote, the rabbit or the badger, is something farmers can do something about, the accuracy of their aim, the legality of their actions, or the disputed evidence aside. The letter boxes are more symbolic of the underlying factors that also, indirectly, affect animal welfare through institutional and social structures.

The problem is also wicked because underlying poor animal welfare is, in part, a failure of a system, of institutional and social structures. Presently, and for much of the last 10,000 years, most of us obtain our food and economic livelihoods from farming. Agriculture is both a business and a way of life, not just to those directly farming, but to humanity in general, providing or supporting the opportunity to build societies and civilizations. Yet this system has its failures, for example:

- farmers unable to borrow money to obtain stock food;
- dairy cows burnt out by the metabolic demands of producing vast amounts of milk;
- farmers paid as little as 1/5th of a penny for each broiler or meat chicken, 25 pence for a sheep;
- animal advocates reverting to covert video footage and television exposés to get improvements in animal welfare;
- an expectation that famers could conduct their own animal welfare assessments to save time and costs;
- cattle, sheep and deer living in mud in winter feeding paddocks and feedlots;
- people appalled and tormented by images of the violent, physical abuse of young calves;
- stockmen barely able to manage mastitis, lameness and infertility in high-producing dairy herds;
- farm workers fired for using their personal finances to buy medicines for sick animals;
- farm staff having to wear earphones with built-in radios to deal with the screams of piglets being castrated and tail-docked;
- the hypocrisy of consumers who pay as little as possible yet expect their meat to have been produced naturally;
- farmers receiving no thanks for providing consumers with pork;
- farmers having to undergo yet another 'hellishly expensive' farm audit;
- failing to provide a pig with a rubber mattress because the vet who gave the instruction was considered arrogant, giving orders instead of treating the farm worker as an equal;
- animals removed from their dams at birth to ensure dairy production systems are highly efficient;
- a farmer giving up because of the excessive demands of having to keep records for assurance and traceability;
- farmers not implementing advice because they have too little time, resources or finances;
- infertile modern dairy cows, requiring an average of 15 injections just to get pregnant;

- a farmer producing enough food to feed 130 people but struggling to feed four on his farm;
- a fear of white cars, the vehicle of choice of one country's animal welfare inspectors;
- the premium offered to farmers for high-welfare beef being too little to recompense for the change in farm practices;
- young farmers guarded about revealing their occupation for fear of being judged;
- the modern dairy cow feeling hungry, tired, full up and sick at the same time;
- supermarkets making more profit per litre of milk than farmers receive for producing it.

These examples, some published, others reliable anecdotes, may not be common or indicative of all animal husbandry, but they are real for both the animals and the people involved. What these 'failures' do is highlight that animal welfare is not just people failing to care for animals, but there is also something about the system that is making it difficult for them. The underlying reasons are embedded within our societal institutions. Indeed, 'If there were no financial and practicable constraints associated with livestock production then there would be few, if any, animal welfare problems' (Petherick, 2005). Financial and practical constraints may not provide sufficient justification for some practices in the future. It is also helpful to appreciate that things cannot always be resolved by looking directly at an issue. For instance, the viability and acceptability of dairy farming are linked with the transport and meat processing industries, the beef industry in some countries being linked to the grain industry. Further complexity comes in the form of state and non-state intervention, for example, financial support through subsidies – and in the different perceptions people have of their role in animal welfare. For example, veterinarians, who once asserted their ethical superiority over unqualified men, have historically, except for a small number of very influential individuals, played little part in the social development of animal welfare, even being criticized for not leading improvements until recognizing the potential conflicts between the interests of animals, their owners, society and the profession (Woods, 2011a).

Yet society's usual responses to instances of poor animal welfare, namely exposés, regulations, protests, codes of welfare, prosecutions and so forth, fail to acknowledge this complexity. Neither do they acknowledge the tension between the compromises to animals and the benefits to humans, the underlying drivers, the aetiology of the disease. We persist in telling people how to treat farm animals without really acknowledging why they are treating them, or having to treat them, in that way in the first place. To address this, it is necessary to understand the whole system, its history, its dimensions (religious, social, economic, scientific etc.) and its interactions – the complexity of animal welfare.

Few works consider the beginnings and transformations of our relationships with farm animals, or indeed within the wider gambit of our lives. Such an understanding is, arguably, necessary if we are to intervene in a system with social, religious, scientific and economic dimensions, all making animal welfare what it is today, a simple (at least for animals) but complex topic (for society) and fascinating (for the author at least). Only by understanding the complexity, diversity and interdependence of animals and people through the perspectives of history, culture, science, ethics, common sense, animal husbandry, economics and the aims of agriculture, can the relationship between humans and animals be maintained and enhanced fairly and sustainably. The remainder of this book explores the interactions and influence of these sorts of factors, acknowledging the biology of the farming system and appreciating what our culture is doing, and can do for agriculture. While most modern criticisms quite rightly demand a change in farming practices, for example, to enhance animal welfare, the present work aims to understand why those farming practices are the way they are and what we need to do to change them. To paraphrase Ane-Karine Arvesen (1941–2005), a respected Norwegian diplomat reflecting on terrorism, 'we've got to find out why people do this – and what we might do to prevent it. We're not going to stop it by talking about how animals are faring' (Fisk, 2008). These sentiments are reflected in a list of ten ways to lose the animal welfare argument, compiled by a US rancher and animal science academic, David Daley (see Peter, 2010):

- assuming science provides answers
- using economics to justify practices
- assuming the need to defend agriculture
- assuming we can't do better at animal welfare
- attacking those who disagree
- not being willing to listen
- assuming the lunatic fringe is the public
- being reactive rather than proactive
- assuming those who disagree are stupid or evil
- not building coalitions that include consumers and the public.

Complexity presents a challenge for three reasons (Jones, 2011). First, the capacity to tackle the problem is distributed across a range of players. If wider society is a significant beneficiary, is there a need to examine how society bears that responsibility? Although there is a danger in widely apportioning responsibilities – 'the goat which has many owners will be left to die in the sun' – it is not enough to criticize farming, to say what is wrong, to lay responsibility at the feet of others, without understanding fully the reason for the way things are as they are, or understanding that our civilization is based on a complex economy – an adaptive network of dynamic ecological economic, social and political relations (Capra, 2002). Secondly, complexity means problems are difficult to predict, requiring responses to have the flexibility to adapt to emerging issues. The final challenge of complexity is that problems involve conflicting goals, the range of cultural, social, economic, religious and other dimensions making the balancing of human benefits against compromises to animals an art.

To summarize this chapter then, and reflecting the complex ecological and social problems humans are regularly exposed to, animal welfare has different dimensions including:

- the state of the animal (what it feels and how it performs);
- our expectations and prejudices of how it feels and of what it needs;
- our expectations and prejudices of the benefits we get from animals; and
- the multitude of financial, economic, political, cultural, legislative, technological, religious, historical and other influences on the human-animal relationship.

Failing to communicate both the simplicity and complexity, to tell people and involve them in the real nature of animals and farming, risks others, however well-intentioned, informed or ignorant, intervening and exploiting them from their own perspectives and for their own ends. It is largely the product of well-intentioned ignorance, as Wilson (2016) describes conservation science in *Half-Earth*, as having an Anthropocene world-view. Interventions may be valuable in themselves, make people feel good and reflect expectations of care, especially of young and vulnerable animals (Mellor et al., 2010), but do they necessarily address the underlying drivers that justify 'necessary and reasonable' animal suffering? It is the deeper roots that need addressing, the systemic dysfunction that animal welfare is a symptom of, the disconnect between realities and expectations. Similarly, it is noted that many of the livestock industries' environmental problems, whether excessive resource consumption or land degradation, have their roots in land speculation, human population growth, and resource use unrelated to environmental costs, forces unable to be altered by livestock-sector policy changes. Addressing them directly, however, would dramatically change farming (Durning and Brough, 1995).

Outline of *Animal Welfare Science, Husbandry and Ethics*

Like the dog and humans, this book is about relationships, the long relationship between man and animals. It is also a story about the relationship between science, animal husbandry, history, economics and morals and ethics, our culture. About the forces that have changed animal husbandry into something that concerns people, and the way those concerns have materialized and are articulated.

For many of us, our relationship with animals is with dogs, usually as members of the family. In contrast, we are quietly disconnecting ourselves from farm animals, and each generation more so. Food comes from supermarkets and restaurants. Although having minimal relationships with farm animals, we retain an expectation of how they should be treated. This book explores that disconnect, the effects of it and the animal welfare 'industry' that has arisen in response. Where once we could have looked the animal in the eye and could both honour it and eat it (Pollan, 2002), we now only see animals on a plate. And although we are concerned about the plight of animals, we are mostly treating the proximate causes, neglect and failure to provide resources, without considering in any meaningful way why those resources are not being provided.

Once we were like wolves, part of the natural environment. With the development of extended social relationships, we have become collectively more intelligent, more affluent and more specialized but disconnected from the environment in which we live and from which we obtain our sustenance. That disconnect is also evident in the most intensive forms of farming – animals, farms and even some farmers have increasingly become disconnected from the land, the soil and the vegetation. And if some farmers have become disconnected, then the rest of society, consumers and citizens, are even more disconnected. It is this disconnection that leads to the belief that some animal welfare problems are not necessarily caused by farmers and others needlessly exploiting animals, but by society's indirect demands of farming. Farm animals, and farmers, like many others in society, are on a treadmill, and with modern breeding and technology that treadmill is getting faster – animals are producing more and sooner and both they and farmers are working harder. And no one knows how to turn this system off (Atwood, 2008). This book is an attempt to understand that treadmill. As well as 'improving' animals, ensuring they 'run' better and faster, it seems sensible to at least acknowledge, if not address, rather than ignore the underlying reasons. There is something counter-intuitive about making farming perform better without asking why. To some extent this does not matter to an animal that may be suffering, except if animal suffering could be minimized by addressing the forces from beyond the farm gate that impinge on the way animals are treated. Until society, and especially those concerned with the welfare of animals, acknowledge this possibility and investigate the potential for changes from outside the livestock sector, contemporary animal welfare will continue to increasingly unsustainably and unfairly deal with the symptoms rather than the causes by making judgements without knowledge of the context necessary for sustainable and equitable solutions. On one hand it could be argued we are becoming too precious about animals, on the other we are in danger of letting economics become the dominant paradigm at the expense of ethics.

I explore the reasons and history using science, philosophy and anything else, from anthropology to zoology and from common sense to ethics, to understand the disconnect. The aim is to draw into animal welfare a more nuanced understanding that will hopefully produce more equitable and sustainable animal welfare. Understanding how the past bears on the present and the future is crucial and Chapter 2, *Drawing on the wealth of agriculture*, delves into the history of the relationships between farm animals and humans. This narrative of farming and humanity links hunter-gatherers with modern farmers, and the modern elite drawing extensively on the wealth of agriculture.

In Chapter 3, *High farming and hard work*, the impact of modern systems on farm animals and farmers is portrayed – why it is that farmers are intensifying, 'growing two blades of grass where one grew before', often without involving society's 'consent' for practices that seemed sensible but proved contentious. It is the controversial nature of those practices that has brought society's concern throughout much of history, the subject of Chapter 4, *Animal husbandry from beyond the farm gate*. There have been many influences and influencers, critiques and stories, which collectively will change not only farming, but also society's use of animals, a veritable animal welfare industry.

Although there is much concern with animal farming, Chapter 5, *People are people through animals*, reminds us of the important ecological and social connections between animals and people, but that many people are disconnected, living their lives without any appreciation or acknowledgement of the real lives of animals upon which society depends. This disconnect is addressed in Chapter 6, *Thinking like a mountain*, suggesting that sustainable and equitable animal welfare should be based on a greater understanding of civilization's biological foundation. The final chapter, *The fall and rise of the hunter-gatherer*, draws together two ways of living – using the natural world as a resource and being part of that natural world.

Animal Welfare Science, Husbandry and Ethics: The Evolving Story of Our Relationship with Farm Animals is a story, one informed and illustrated by the author's experiences and prejudices, family and colleagues, education and expectations. It is couched in terms of evidence provided by science, a knowledge of husbandry and technology, and morality borne of common sense, a knowledge of ethics, and the incredible imaginative insights available through stories, art, mythology and everything else humans do, to give 'a broader perspective of how an animal's welfare is influenced or determined by his or her location within a complex of human activities' (Anthony, 2009). In other words, not just the animal's biology and the circumstances of its husbandry, but also the economic, social and moral foundations of those who care for animals, and have an interest in how animals are cared for, and the roles animals have in human communities. If the theme of the story is correct, then we need to start having conversations about the way we think about animal welfare, and possibly about reconnecting as a society or community. Collectively, with our biology and values, by expanding the role of citizens, so that we, individually and collectively, have the confidence, resources and opportunities to do what is required and what people want.

Farming, science and ethics – some history and prejudices

After many years immersed in farming, science, animal welfare and ethics, this is my personal reflection. It is important then that you, the reader, understand something about me and the way I have come to frame animal welfare in this book by acknowledging its various and complex dimensions. My story is told from the privilege and prejudice of having grown up on a family sheep and beef cattle farm, of studying, working in, and meeting people in animal science around the world, becoming interested in philosophy and ethics, serving on several national advisory bodies and, most recently, in the public service. The bibliographic *Further reading* sections at the end of each chapter give further insight into some of the works that I have drawn on and that have shaped my viewpoint. It is a journey using insights from different sources and times – it is not an appraisal of the current status of farm animal welfare. Furthermore, many of the examples of poor welfare highlighted are, and have been, the focus of scientific, veterinary, industry, animal advocacy and farmer led improvements.

Some of the aspects have been experienced or witnessed first-hand. For example, the depopulation of rural communities – I was one of 52 children at the local primary school, my nephews were among just 8 children at the same school. Continuing the tradition stretching back to at least the Athenian statesman, lawmaker and poet Solon (c. 630–c. 560 BC), I was advised to 'get a trade first' if I wanted to go into farming. And growing up at that time, poultry meat was reserved for special events like Christmas or New Year and involved wringing hens' necks and plucking them. In later years, I was part of a study of the impact of helicopter hunting on deer numbers in NZ's largest National Park; while studying animal science in England I lived near where the early livestock breeder Robert Bakewell once farmed; and, on returning to NZ, I undertook research on the control of reproduction in red deer through the generous assistance of one of the pioneers of deer recovery and farming, Sir Tim Wallis.

My starting position is that we cannot live without having an impact on animals, and that for most of us, our food and livelihoods are derived from compromising animals in some way. In doing so, however, we have a duty to give animals a good life and a quick and painless death. Mankind and animals are inextricably linked, and it is the quality of this relationship that is of most interest. While the immediate causes of poor animal welfare reflect the relationship between the animal and the stockman, some of the ultimate causes reflect the relationship between society/culture and the stockman, captured by the Irish proverb, 'If I am a gentleman and you are a gentleman, who will milk the cow?' We cannot really maintain and enhance relationships by ignoring them, their fullness and complexity, or our interdependence. Things rarely change until we combine reason, emotions and stories. It is unwise, however tempting, to become fixed on simple and direct, even technical, solutions to complex problems, especially when they have important economic, political, social and other dimensions. I believe in farming and using animals, and that most people are trying to do the right thing, although some are pushing, and some are being pushed beyond, the limits of society's level of acceptability. Furthermore, most people caring for animals are not evil but working to the best of the circumstances they find themselves in. I also believe we can work through this together but am unsure whether we will do in our distracted and disconnected age characterized by people responding to the immediately obvious without seeing, or at least being provided with, the history, context and rationale for animal welfare compromises.

What most concerns me is that the costs and benefits of using animals are not necessarily understood and fairly shared. Over many years, I have asked audiences what is the worst thing that we do to animals (inevitably neglecting them or confining them) and to farmers (bombard them with information and treat them as if they were all the same). My insight is that 'once you see the forces that govern behaviour, it is harder to blame the behaver' (Wright, 1994). I also do not believe we can continue to increase animal production and address the concerns, particularly animal welfare, associated with intensification. That leads to concern about what society does when we reach the limits of animal welfare, animal husbandry and environmental resilience. I do not believe labelling and premiums will ensure consumers pay a fair price for animal welfare. Nor do I believe science and technology can provide all the answers, as they do not provide the values that are the underlying drivers of intensification. I do not believe values and ethics will provide an answer either, without incorporating them into economics in some novel, innovative or meaningful way, perhaps requiring that we better utilize all our ways of knowing, something that modern society struggles with. How can society's economic system, built on growth, respond when its resources are finite? The story is one I have pieced together and hopefully it will help the conversations I believe we need to have to appreciate our social and ecological dependence. As Atwood (2008) believes, that power has shifted from those who farm, to those who process, and to those who market, retail and govern. And as farmers, processors, marketers, retailers and those who govern have different values, interests, emotions, experiences, stories and expectations, there is a need to ensure they are joined.

However, perhaps the most important thing to be aware of is that this story has a strong bias towards farm animal welfare in Europe and North America, as well as New Zealand, partly reflecting the author's experiences. It is also deliberate; modern concerns with farm animal welfare mainly reflect the development of intensive, industrial systems characterized by confined and barren environments most associated with western, educated, industrial, rich and democratic countries. While intensive, industrial systems have spread from North America and Europe where they have arguably been most developed, so the future of animal welfare may well lie in appreciating the full diversity of humanity as studies of human behaviour and psychology have begun to acknowledge (Henrich et al., 2010a,b). I return to this in Chapter 6.

Following the UK's Farm Animal Welfare Committee (2016), 'farmer' means a person who cares for livestock on a day-to-day basis, and anyone behind the farm gate, that is, the farmer, his or her family, farm workers and employees. Interestingly, the term derives from the French *fermier*, a progressive peasant who, in the Middle Ages, was eager to produce for commercial markets through renting land. If he could get land, the product of his labour was his own. Farmers were thus landless people seeking land for the value it imparted to their labour (Strange, 1988). The word was not common till the 18th century as most people were engaged in farming and so it had little descriptive use, most farmers being classified according to status and other occupations. Also, a couple of points about language. I use 'man' in the sense of human beings in general, that is, mankind or humans, recognizing that this includes men, women and children,

Figure 1.7 The author, at two years of age, showing a pup at a farm-dog trial (*Photo: Kandid Kamera Kraft*).

all of whom have played and continue to play significant roles in farming and in animal welfare.

This chapter began with a quote from my childhood. I am ending it with an image (Figure 1.7) from the same period. It is of a 'pup show', a regular feature at dog trials where humans and sheepdogs compete in demonstrating the skill of the human-animal bond in moving sheep, an important part of animal husbandry remembered in the bronze statue at the beginning of the chapter. This image captures the attraction we have for animals from a young age, even though animals might not always see it that way, or at least agree with the techniques we use to ensure animals 'fit' our worlds (it was probably the first time the pup was restrained by a lead). It also symbolizes the unequal nature of the relationship.

Further reading

How the Dog Became the Dog: From Wolves to Our Best Friends. **Mark Derr**. Scribe, Melbourne, Australia, 2012.

The Intimate Bond: How Animals Shaped Human History. **Brian Fagan**. Bloomsbury, New York, NY, 2015.

Assessing animal welfare at the farm and group level: the interplay of science and values. **David Fraser**. Animal Welfare 12 (2003), 433–443.

The Welfare of Dogs. **Kevin Stafford**, Springer, Dordrecht, the Netherlands, 2006.

Ethics and animal welfare: the inextricable connection. **Jerrold Tannenbaum**. Journal of the American Veterinary Medical Association 198 (1991), 1360–1376.

CHAPTER TWO

Drawing on the wealth of agriculture

History makes little sense without prehistory, and prehistory makes little sense without biology. (Wilson, 2014)

Not far from the bronze statue of the dog and the Church of the Good Shepherd lies a golden tussock-covered area of the NZ high country. It was here in 1871 that early European settlers released seven red deer from Scotland, the founding stock of the region's world-renowned herd. NZ had few native land mammals and the release aimed to provide a resource for hunting for food and sport. As well as traditional farm animals, possums, wallabies, hares, rabbits, mice, rats and mustelids were also introduced along with, unsuccessfully, llama, wildebeest and zebra (Druett, 1983). Extant throughout much of Europe, North Africa and Asia, red deer have long been hunted, appearing in various archaeological sites. For example, excavations of bone accumulations alongside hearths and Stone Age artefacts in Kebara cave on Mt Carmel in Israel, reveal Neanderthals hunted two main larger mammals, mountain gazelle and Persian fallow deer, as well as red deer, aurochs and wild boar (Speth, 2006). In the UK, Neanderthals occupying caves included red deer in their art (Figure 2.1), along with bison, horses and birds.

Deer have a long history of being managed in parks and enclosures in antiquity in many parts of the world. And, like many species, during the settling and expansion of colonial countries, red deer were introduced into Australia and New Zealand. From the middle of the 19th century, there were perhaps as many as 300 liberations of deer in NZ from parks in Britain and Australia. Red deer found the NZ environment, climate and vegetation, which evolved in the absence of herbivores except for the now extinct flightless moa, and lack of competition and natural predators to their liking, and dispersed and thrived. Populations erupted and progressively depleted preferred plant species. Concerns that deer were displacing livestock and damaging pastures, exotic forests and crops led to culling,

Figure 2.1 A representation of a shallow engraving of a red deer stag on the sandstone wall of Church Hole Cave at Creswell Crags in central England from 13,000–15,000 years ago (Pike et al., 2005). Caves and rock shelters in the limestone outcrops in the gorge were used by hunter-gatherers, Neanderthals and early modern humans, from 55,000 years ago.

initially undertaken to remove poor-quality animals and preserve trophy heads, but later to preserve the integrity of vegetation in high-rainfall catchments prone to erosion, and finally to protect the native vegetation (Caughley, 1988). By 1930, a 'The Deer Menace' conference sanctioned culling and government shooters, supplied with provisions dropped from fixed-wing aircraft in difficult country, killed a lot of animals. Deer hunting, immortalized in Barry Crump's *A Good Keen Man* (1960), became part of the nation's masculine psyche.

Initially, markets for venison were serviced by hunters on foot near road ends, off rivers serviced by jet boat, and by fixed-wing aircraft from improvised strips. Eventually, helicopters provided access to the more rugged and remote parts of the landscape, resulting in a huge reduction in deer numbers. Commercial meat recovery, hastened by the securing of a market in the US, became a multimillion-dollar business by the 1960s and 1970s, the hunting so effective and sustained that there was little need to control deer numbers. As numbers in the wild reduced, deer farming began. Animals once shot, were lured into paddock traps and corrals, captured with the aid of tranquilizers and bulldogged or netted from helicopters. Throughout this period, there was remarkable innovation and skill, entrepreneurship, fierce competition, people who were prepared to push machines to the limit, poaching and the inevitable bureaucratic interventions. Legislation dictating that wildlife could not be privately owned had to be revised. Deer once protected by law were then hunted and finally became a resource for commercial venison recovery and eventually farming. The success of NZ's deer farming industry owes much to the abundant, readily available and free-ranging wild population of red deer, and the vision, skills and sheer guts of pioneering hunters, helicopter operators, farmers and others. It also owes much to technologies as diverse as net-guns, custom-made fencing, hydraulic crushes and animal health interventions (Caughley, 1983; Challies, 1985).

It has now been 50 years since deer farming was first legitimized in NZ and elsewhere, most notably Scotland (Blaxter et al., 1974). In that short period of time, to paraphrase Masanobu Fukuoka (1913–2008), a Japanese farmer and philosopher, in *The One-Straw Revolution* (1978), if you change the way you grow and manage the animal, then you change the animal. The changes red deer have been exposed to have been remarkable, the selective pressures on the population of wild deer over such a short period huge. Hunting selected against animals wary of man and machine, and nearly a fifth of animals captured succumbed to stress or post-capture myopathy, a complex disease associated with strenuous exercise and capture resulting in metabolic breakdown and severe muscle damage (Wilson, 2002). Captured animals had to cope with a farm environment that, at times, did not respect their biology. For example, the motivation for isolation around the time of giving birth (Arman, 1974; Arman et al., 1978; Cowie et al., 1985) can be thwarted by farm fences, resulting in an almost stereotypy-like repetitive behaviour whereby hinds pace a fence-line without resting or grazing (Figure 2.2).

Similarly, human disturbance at calving resulted in an unusually high number of newborn deaths when deer were first farmed (Kelly and Whateley, 1975). Finally, with breeding and nutrition characteristic of modern farming, the animals are now bigger, mature earlier and produce more. They generally also inhabit a less varied environment, are kept at higher densities, are exposed to normal farm management practices that reduce disease and ill-health and can be prevented from behaviours such as wallowing and mate selection (Fisher and Bryant, 1993). Interestingly, the pioneers of deer farming discussed purposefully interbreeding their farmed stock with wild animals to maintain the vigour of farmed animals.

There are now around a million deer on 2000 farms in NZ with wild populations still found throughout the country, hunted by recreational hunters and managed or controlled by authorities to protect sensitive environments. This relatively recent adaptation from the wild to a farming environment has been accompanied by the resolution of many issues, such as behavioural problems at calving. However, it has also brought new ones associated with increased productivity. Better nutrition and growth, and

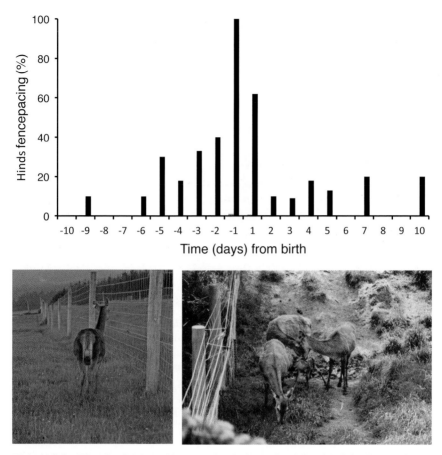

Figure 2.2 The distribution of fence-pacing in farmed red deer hinds in very small paddocks leading up to calving where they were unable to seek isolation. In the wild, hinds usually calve in isolation, returning to the herd after one to two weeks, a behaviour also apparent in other animals including extensively farmed beef cattle and sheep, and wild and feral pigs. Fence-pacing, first evident as gentle walking (lower left), increasing in pace and intensity immediately before or after calving, may be a sign of frustration with being unable to seek isolation during parturition. The lack of isolation may also result in interference from other hinds during the time that hind and calf normally bond in isolation from other deer (*Photos: Mark Fisher, left, and Jens Sigsgaard, right*).

selection for heavier velvet antler has seen spiker heads (traditionally stags in their first season, known as spikers, developed a single spiked antler) transformed into heavier, mature-like heads (up to 30 or more points), raising questions concerning the ability of young animals to hold their heads up comfortably. Similarly, hind reproduction is being manipulated to make deer farming more efficient (Figure 2.3).

The domestication of red deer occurred within a relatively short period, facilitated by 20th-century knowledge, technology and markets. However,

Figure 2.3 A modern farmed red deer hind (*Photo: Mark Fisher*). Note that this hind is heavily pregnant, in this instance with a hybrid Canadian wapiti x red calf, part of work to find the most efficient system of production, nominally a large male mated to a small female. Note also that the pasture in the foreground has gone to seed. Deer, being browsing species, calve in early summer after the spring peak of farm pasture production, necessitating research to induce earlier mating and thus calving, and thus more efficient use of pasture for lactation. (Available in colour in the plate section between pages 178–179.)

the attributes and principles of this modern example of domestication generally resemble those of more traditionally domesticated farm animals, part of a dynamic process that began in Neolithic times as part of the transition of humans from nomadic and egalitarian hunter-gatherers to the materialistic, hierarchical societies of modern civilization. Although there are many ways to describe the changing history of our relationship with farm animals, they are portrayed in this chapter in terms of Paleolithic hunter-gathering, Neolithic farming, the Agricultural Revolution of the Middle Ages, and finally the Industrial and modern eras (Table 2.1). The changes outlined in these general periods track the increase in energy available throughout history. Whilst, with the benefit of hindsight, changes in agriculture are often referred to as revolutions, these periods were likely a continuum, with ideas and practices in one era becoming more prevalent and important in a subsequent era. While there are many very detailed works on the beginnings of farming it is necessary to provide a very general description of the changes to both animals and people, and somewhat neglecting plants and landscapes. However, it should be acknowledged that

Table 2.1 *The approximate stages of agricultural development, the increase in human population, and the sources of energy available (kcal/day) at different stages of human history (from Mazoyer and Roudart, 2006; Simmons, 1995).*

Age	Communities (total population)	Energy source and per capita availability
Paleolithic (40,000–10,000 years ago)	Hunter-gatherers (10 million)	Plants, animals and fire (*c.*3000 kcal/day)
Neolithic agricultural revolution (10,000–5000 years ago) and early agricultural societies (5000–1000 years ago)	Farming (50–250 million)	Plants, animals and fire (12,000–26,000 kcal/day)
Agricultural revolution of the Middle Ages (from 1000 years ago)	Industrial (250–500 million)	As above plus coal, oil, natural gas and hydro-power (77,000 kcal/day)
Agricultural revolutions of modern times	Modern (500–6000 million)	As above plus nuclear power and alternative sources (230,000 kcal/day)

there are many intricacies and complexities – different regions, landscapes, flora and fauna, climatic changes, resources, populations, times, cultures and explanations – worthy of, if not demanding, critical attention. The changes are neither revolutionary but gradual and long-lived, nor linear. As people needed to adapt to changing circumstances, they undoubtedly adopted techniques and practices that may have been prevalent in earlier times and that contributed to developments in subsequent ages. Hunter-gatherers farmed, and farmers hunted and gathered, in response to their respective needs.

Prehistory, hunter-gatherers and proto-farming

All animals are equal.

Prehistory is the time before words were recorded, covering the period from the origin of humans 6 million or more years ago, to the use of stone tools in the Paleolithic or old Stone Age 2.5 million years ago to the beginnings of agriculture in the Neolithic or new Stone Age 10,000 years ago (Mithen, 1996). Until then, most humans were foragers, gathering, collecting, scavenging, fishing and hunting or hunter-gatherers. Living in small, closely knit family bands or community groups, they got most of their food from wild animals and plants. The nomadic lifestyle was not normally especially arduous or uncomfortable if they never remained in one settlement for very long but kept moving to ensure local resources were not exhausted. Although there were food shortages and starvation, hunter-gatherers rarely exceeded the carrying capacity of their environment. They were probably highly organized, had much leisure time (modern hunter-gatherers spend little time hunting and feeding, see Serpell, 1986) and, because they were mobile, had few possessions. There was some division of labour – men hunted and women gathered – but everyone lived as equals.

Hunter-gatherers, what we have been for most of history since diverging from the ancestors of our living apes 7 million years ago, probably saw nature and the natural world as if it was a social being, a single world of people, animals, plants and the land and water. The change from a hunter-gatherer to farming required four things:

- the ability to develop tools for harvesting and processing plants, and the propensity to
- use animals and plants to acquire social prestige and power,
- develop social relations with plants and animals like those developed with people, and
- manipulate plants and animals.

These features, critical to the rise of farming, required a modern mind. Having a theory of mind for other animals, being able to think like them, may have been of great advantage to hunter-gatherers, enabling them to predict animal behaviour and develop sophisticated hunting techniques. Anthropomorphism and totemism may also have similar origins. But most important was the ability to empathize with animals. Along with care, empathy is arguably one of the most important features of modern stockmanship (Bradshaw and Paul, 2010; Gatward, 2001).

In *The Prehistory of the Human Mind*, Mithen (1996) argues that a critical step in human evolution was the rise of the flexible mind through the emergence of cognitive fluidity. Knowledge, thoughts and experiences gained in different and discrete domains or areas of specialized intelligence (general, natural history, social and technical intelligence, together with language) could flow freely around the mind resulting in 'an almost limitless capacity for imagination'. It was this mind that enabled man to begin to dominate animals and would eventually enable modern humans to develop science and medicine, turn food into a commodity and a source of wealth and power, create empires and religions that would spread throughout the world, and put men on the moon. The result of some unknown structural and/or cultural development during the Old Stone Age, modern man began to see the natural world in a new way. The evolution of the mind, language, the beginning of a separation from nature, the design of tools, the creation of art and the development of religious beliefs brought problem solving and innovation. Reflecting that more creative mind, traces of art, rituals and new technologies appeared. Mankind began to develop social relationships with animals and plants and to use and manipulate them to acquire social prestige and power.

These changes, in the mind, cognitive fluidity and symbolic thought or language, laid down in the Stone Age, lie at the root of humanity in the modern world. Human uniqueness lies in these changes resulting in greater social cooperation and the ability to empathize, rather than being more competitive and selfish. The cause of the changes is not known: possibly a mutation, possibly an effect of gradual evolutionary change, possibly the result of interbreeding with Neanderthals. While humans were anatomically and behaviourally modern around 40,000 years ago, it took a set of environmental and economic circumstances to provide the final spur for farming: the dramatic climate changes at the end of the last ice age and the need to feed relatively high local populations. Greater powers of innovation and cultural changes at a time of retreating glaciers provided new ecological opportunities for animals favoured with the ability to adapt, and for man.

Generalized hunter-gatherers had general tools, but, in response to environmental and population pressures, Stone Age man was driven by a need to exploit the environment in increasingly more efficient ways and became specialized hunters and skilled craftsmen. Hand-axes, scrapers, points, blades, sharp-pointed borers, burins or chisels, bone and antler tools were developed, as well as art and other ornaments. In turn, those technologies would have been reinforced by increasing experience and knowledge of the habits of animals, and of the easiest ways to capture them, in turn resulting in more specialization. The specialization was cultural as well as biological – hunters acted on the environment and the animals conditioned the technology, the behaviour, the social organization and the thoughts of hunters. They improved by deepening their understanding of their environment and the relationships they had with prey animals (Orquera, 1984). Efficient hunting was complemented by efficient gathering, eventually cultivation, and the specialization of work within the community. The change from everyone doing everything to a complementary division of tasks within groups of people enabled further specialization and it was this, and the exchange of products and services between different groups of people, that would become the source of economic prosperity.

As human numbers and hunting increased, some animals retreated and some, like goats and pigs, remained close but in difficult country. Corrals made of wooden posts, brush, stone and earthen walls or box canyons would have been used to hold animals as a supply of fresh meat. Pigs and dogs were scavengers, domesticated when human camps or settlements became an important food source. Once in place, this specialization process reached a point of no return – food production had to increase as the human population increased as hunter-gatherers exploiting their habitat or range came up against other growing populations of sedentary hunter-gatherers. The effort in tilling, sowing, weeding, protecting and harvesting cereals was not easily reconciled with a nomadic hunting and gathering lifestyle. Nevertheless, farming began to have an enormous impact on the ecological success of those who practised it, penalizing humans and animals who were not part of it. Farming may not have been adopted by choice – people may have been forced into it as their hunting and gathering lifestyle became more difficult. Less dense populations of hunter-gatherers were displaced or assimilated by the might of denser populations of farmers.

Processes like farming that appear in one period inevitably have earlier beginnings, humans modifying the environment well before the beginnings of agriculture. Hunter-gatherers in Africa fished, hunted water-fowl and other birds as well as a range of small and medium-sized game, especially Barbary sheep. But they also lived with animals, experimenting with the herding of wild Barbary sheep, including stalling and feeding them in caves when feed and water were scarce, a sort of proto-farming existing since the rise of modern humans. Wild animals would have been managed with practices such as 'firestick farming', not only flushing out prey but also spurring regrowth of vegetation leading to higher densities of food-bearing plants preferred by animals. Fire also cleared ground for human habitats, facilitated travel, killed vermin and was used in warfare.

The sedentary lifestyle associated with abundant resources for hunting and gathering allowed a material culture to develop and humans began to accumulate wealth for the first time. In turn, this enabled non-productive elites to form, not because they were needed but just because they could extract resources from others; hunter-gatherers would not have had much use for wannabe 'big men', people who tried to rule or manage. However, when food was stored it provided an opportunity for others to steal and the elites, those who live on the productive work of others, came into existence. Interestingly, some hunter-gatherers chose to curb elites by growing root crops that rotted if not harvested, ensuring they could not be stolen (Cochran and Harpending, 2009). One of the earliest groups of socially and technologically advanced modern humans were of the Gravettian culture, known for the site in south-western France where it was first recognized. These people lived across the western Eurasian Plain and into the Mediterranean from 30,000 to 22,000 years ago. They constructed their own shelters, hunted reindeer, horse and hare, scavenged the remains of mammoths and organized hunts requiring ingenuity, courage and cooperation as well as an intimate knowledge of the terrain. Their villages may have been community focal points. And they invented an economy in which surplus food was produced and stored in pits dug into the permafrost, necessitating its protection. The Gravettians may have begun to divide their labour, develop small and portable tools, and partly give up their nomadic hunter-gatherer lifestyle. There is also evidence of art, textiles and basketry, and net-making for both fishing and trapping small animals. They brought together many of the cultural, technological and social skills seen in farming communities.

Another group of people, the Natufians, were precursors to humanity's earliest farmers, about 15,000 years ago in the eastern Mediterranean when there was a warming of global temperatures. The Natufians had a material culture that included houses, elaborate burials, and technology such as flint sickles and large mortars for gathering and grinding wild plants and seeds. Living in an ecologically rich area of the Mediterranean and surrounded by animals and plants such as gazelles and cereals, they also settled permanently although initially without farming, relying on hunting, fishing and gathering. In the Natufians were the beginnings of a cultural revolution, human settlements becoming larger and more complex, and the people more demanding of the environment. Fire may have been used to enhance grazing for gazelles, an important hunted animal. As the climate subsequently changed again, becoming colder and drier, foraging became more difficult and the Natufians moved, diversified, or intensified their system, becoming more reliant on cultivating plants. Most interestingly from the perspective of the underlying theme of this book, the Natufians exploited their resources more intensively. Although harvesting grains by beating is quicker and easier, harvesting with sickles resulted in a better yield. This is one of the earliest examples of intensification; the need to exploit a limited area of land meant that it was more efficient to harvest as much as possible instead of spending more time, and travel, harvesting from a greater area (Belfer-Cohen, 1991). During this time, increased trading occurred over longer distances and status and decision-making systems may have changed from being kin-based to community-wide. As people intervened in nature and produced food for themselves, they began to question their place in nature and the universe. From being spectators of nature's cycles, they began to master and transform nature.

Like chimpanzees and gorillas, humans, as nomadic hunter-gatherers, lived in bands without economic specialization; all individuals would have been expected to forage on land used jointly by the group. There were no formal social mechanisms for resolving conflicts, no formal social stratification and leadership reflected an individual's personality, intelligence and other qualities. The organization of the band was considered 'egalitarian', somewhat like George Orwell's *Animal Farm* (1945) after the revolution, where, 'All animals are equal.' The remainder of this chapter traces the history of farming through several 'revolutions', their path uncannily like the plot of *Animal Farm* highlighted in the selection of quotes at the start of each section. In addition to reflecting on the Russian revolution, there are several other themes in *Animal Farm*, for example, the loss of the agrarian way of life and the impact of mechanization (McCorry, 2013). However, the reason for drawing on *Animal Farm* is that the history of farming reflects the tension between farmers and their environment, and between farmers and others (urban civilization). It is those tensions that farmers and others have sought to lessen with access to technology and knowledge, political influence and, at times, uprisings. Like the rebellion or overthrow of Jones in *Animal Farm*, they are revolutions: the Neolithic, the Agricultural and the Industrial. Unlike *Animal Farm*, the changes are gradual, the term revolution a convenient description from the perspective of history. Although *Animal Farm* may have been a parody of communism, its value here is more in keeping with its original context. Indeed, there is a wonderful similarity between Orwell's pigs moving into Jones' house and occupying the bedroom, and an account of pigs kept in a derelict house in Britain during World War II. While part of the house had collapsed, the pigs had access to the staircase, chasing each other up and down during daylight, and competing for the upstairs bedroom at night – 'I never had pigs do better than that lot' (see Harrison, 1964).

The first revolution is the Neolithic or New Stone Age. It was the time when animals were first domesticated and marked the beginnings of large-scale and permanent agriculture: the growth of cities and civilization. The need to defend land, coordinate activities, trade and store food surpluses, required different rules: ownership, social hierarchies, gods and morals. While it is biologically sensible for hunter-gatherers to be lazy, farmers had to work harder, and selection probably favoured those who did. Farming had few advantages over the hunter-gatherer lifestyle except that it provided more calories per unit space. Its adoption led to a decline in the quality of the diet. But it was the only possible response to sedentary populations of hunter-gatherers having reached the carrying capacity of their environment. Though not adopted by all hunter-gatherers, farming slowly took hold. As populations of hunter-gatherers became more settled, the limitations imposed by the mobility of a nomadic life decreased. People gathered more possessions and population numbers increased. Exacerbated by climate change, the factors for its adoption also included a decline in the availability of wild foods (Flannery, 1994), an increasing availability of domesticated plants and better methods for collecting, processing and storing foods. As prey and plant species were exploited, it became more economic to cultivate and breed them, an increasing reliance on cattle herding one of the signs of intensification. And increases in food production saw further increases in human population density.

Domestication: the Neolithic agricultural revolution

All animals are equal. But some animals are more equal than others.

Charles Darwin stated in *On the Origin of the Species* (1859) that there is nothing easier than to tame an animal and few things more difficult than to breed them under confinement. So began domestication, the modification and management of wild animals making them more useful to humans. Darwin also observed that there was more variation with varieties of domesticated species than within wild or natural populations, both animals and plants. Species that have been domesticated may have more inherent variation, upon which people and environments may have acted. The climate, the interglacial periods of retreating glaciers and warming at the end of the last ice age, immediately preceding the Neolithic revolution, may have favoured animals more readily able to adapt to changing environments (Coppinger and Smith, 1983). This included traits such as the ability to be tamed, which may have begun with women nursing young animals (see Figure 1.6), and to become interdependent and less specialized. Such traits were recognized in younger animals more than in older ones, for example, care-soliciting behaviour, and gave some animals a selective advantage over those without them.

Domestication, then, far from being a human or artificial invention, is a successful evolutionary strategy adopted by animals, including humans, and plants. Success is evident in the numbers of domesticated dogs, goats, sheep, cattle and poultry for example. Those species unable to tolerate humans in close proximity and benefit from help with breeding and birth, finding food and shelter, or protection from predators and treatment for ill-health, were less fit, their eventually threatened populations including the world's most iconic wildlife species, for example, panda, kiwi, bald eagle, tiger, rhinoceros. It is a relationship between one group of organisms and another, a co-dependent relationship usually characterized by the more dominant group, man, manipulating the behaviour, social lives, growth, territory, reproduction and ultimately the fate of animals. From an animal's perspective, domestication is a condition where man controls the feeding, care and breeding of animals, resulting in changes in biology (behaviour, physiology and morphology). It is an evolutionary process resulting from different selection pressures created by that environment along with a lessening of the selection pressures (e.g. competition, predation) found or encountered in the wild, resulting in changes in the animals (e.g. behavioural, physiological and morphological). By changing the environment, man could accelerate the development of traits of importance. Increasingly, humans intervened in the relationship and gained the upper hand by being able to learn and transmit cultural behaviour rapidly (e.g. manipulating the behaviour and reproduction of animals). The uniqueness, importance and power of the relationship come from this cultural component (Zeder, 2006) and its importance is seen in the economies founded on the resources animals provided.

In Neolithic times, the selection of animals must have begun with the use of suitable animals only. Not all species can be domesticated: only those able to cope, thrive and reproduce would be included, the timid and aggressive excluded. Less sensitive, less nervous, less forceful and smaller animals were all typical of primitive domestic animal populations.

In addition to being able to tolerate man, domesticated animals would have needed (Diamond, 2002) to:

- have a temperament making them able to adapt to closer confinement,
- have forms of communication and interdependence and
- breed freely in such circumstances.

Interestingly, modern selection for tameness in foxes (no aggression or fear responses to man) has seen greater reproductive success, in terms of earlier mating during the breeding season and larger litters among those that produce litters (Belyaev and Trut, 1975). Animals destined to become domesticated would also have to have been:

- not territorial;
- living in large wide-ranging herds of mixed sexes organized in hierarchies (enabling greater numbers of animals to live more closely);
- eating many different plants;
- displaying a relatively slow response to danger; and
- adaptable to a social structure enabling them to live under human leadership, dominance or stewardship depending on their biology (Diamond, 2002).

Conversely, animal species that could not easily be provided with a suitable diet or adapt to enclosures, those that grew slowly, were reluctant to breed or had a long interval between births, or were aggressive and lacked dominance hierarchies, could not be domesticated.

These features highlight the importance of considering animals in terms of social relationships. Being gregarious, herd animals would have existed alongside man's early crops. Humans had to learn to handle them while the animals took advantage of the association with man, for example, consuming crop stubble, changing the social relationship to a symbiotic one. Species with a larger proportion of individuals able to adjust, that is, tameable, were more likely candidates for domestication, providing humans were sufficiently motivated and able to make use of them. Having the skills and resources to grow or at least provide enough fodder for animals, and contain them, required additional energy that bands of humans could not have afforded to divert while living on a subsistence diet.

There are several different sites in the world where humans, over thousands of years, developed cultivation and raised animals by making use of materials, social changes and knowledge. The animals domesticated in the main centres of farming were:

- cattle, sheep, goats, pigs, pigeons and donkeys in the Near Eastern centre (Syria and Palestine, commonly the Fertile Crescent) 10,000–9000 years ago but also in India (cattle and sheep), Europe and Asia (pigs);
- turkeys and ducks in the Central America centre (southern Mexico) 9000–4000 years ago;
- hens, pigs, cattle and possibly the dog in the Chinese centre (northern China expanding to the northeast and southeast) 8500–6000 years ago;
- pigs (possibly) in New Guinea about 10,000 years ago; and
- llamas, alpacas and guinea pigs in the South America centre (Peru and Ecuadorian Andes) over 6000 years ago.

Although this occurred at several places, the eastern Mediterranean or Near East, known as the Fertile Crescent, the Levant and upper Mesopotamia, is one of the oldest and most well-known and understood. It was a region rich in resources. It was from here that wheat barley, peas, flax and lentils, as well as the goat, sheep, pig, cow and donkey were domesticated.

Domestication, part of a continuation of a long tradition of using animals, progressed from hunting, which required knowledge of the animals, along with skills, techniques and attitudes towards them. Stalking, ambushing, driving, trailing, trapping and fishing not only required different woodcraft skills, knowledge of animal behaviour, tracking and so forth, but also differences in patience, self-control, courage and ingenuity. Domestication had both animal and human components (Downs, 1960) including:

(1) An initial stage of loose ties between man and animals with the latter freely breeding.
(2) Subjugation of animals and making them dependent on man, with captive breeding resulting in distinct traits, e.g. colour, size, musculature.
(3) The intentional development of desired traits, e.g. large size, with some interbreeding with animals in the wild.
(4) An increased focus on economic (meat, milk, wool etc.) and morphological (horns, colour etc.) traits.
(5) Significant differences in traits becoming evident meaning that interbreeding with wild types would have spoilt the desired traits.

As well as this animal-centred or zoological understanding, there were human-centred or sociocultural/anthropological criteria:

(1) A need or use for controlling, protecting and breeding animals, e.g. milk, meat or eggs, draught power, social prestige or religious sacrifice. In other words, the relationship between humans and the animals had to change. In modern times, making our economic livelihoods from animals is an accepted use and many people benefit from processing, marketing and retailing domesticated animal products, with social institutions and societies organized around them.
(2) Controlling animals to make use of them. For example, where once chickens roamed farmyards, orchards and fields and laid their eggs in nests, many are now confined to sheds and cages, their conditions determined by human requirements for plentiful and inexpensive eggs.
(3) Nurturing the animals, protecting them from disease and parasitism, and providing sufficient food, water and shelter proportional to the amount of control imposed (e.g. sometimes preventing them from obtaining their own resources). The provision of conserved forage, hay, silage etc., during periods of slower forage growth such as in the cold of winter or during droughts, is a prime way of overcoming the ecological constraints limiting wild populations of animals. It is most marked in modern intensive systems where the control over animals precludes them foraging for themselves, being reliant on man for all their resources.
(4) Human involvement in animal selection and breeding. At first, it

would have been involuntary; only those individuals and species amenable to being in close association, perhaps captive, with man and that fulfilled a need, became domesticated. And man had to learn how to husband them, especially large and dangerous species. Later, some animals would have been selectively killed, and eventually advantageous and desired individuals or phenotypes and traits or genotypes would have been deliberately selected for.

Domestication would therefore have required a degree of social organization, and technology allowing diversion of energy from direct gathering or production of food to experiment and being sedentary enough to remain with the animals until the taming process and breeding produced controllable animals. In other words, domestication is not only biological (taming and managing) but requires technical knowledge and economic, marketing, environmental, political and regulatory support. Interestingly, the main reason why farming new species of fish has failed in modern times is overly optimistic estimates of market demand (Teletchea and Fontaine, 2014).

Early Neolithic man was probably only interested in keeping animals subjugated, making them docile and using them. Providing for animals' needs was probably not considered and their health and welfare probably suffered, animals becoming smaller, weaker and more docile. Early evidence of compromises to animals was evident at some Neolithic sites farming cereals and legumes and herding sheep and goats: pathologies in the foot bones of goats possibly indicative of unsuitable stalling (Crabtree, 1993).

Compared with their wild ancestors, there is little doubt that domesticated animals have differences in morphological, physiological or behavioural traits (Bökönyi, 1974; Clutton-Brock, 1999; Hemmer, 1990; Price, 1984; Setchell, 1992; Zeuner, 1963). There are many examples:

- body size has increased in the horse and hen and decreased in the sheep;
- a greater diversity in horns in sheep and cattle where selection for religious or aesthetic purposes often produced fantastic-sized and -shaped horns;
- many farm species have developed a piebald coat pattern, whereas, except for the hyena, it is rare in the wild;
- an increase in the number of tail vertebrae in sheep from 10 in wild to 13–35 in domesticated animals, where the tail is often fatter, and abnormal twisting of the tail in the pig;
- smaller skulls relative to their bodies, the loss in brain size seen

mainly in changes in the centres of sensual perception and perhaps a lack of perception;
- increased fat and muscle deposition in sheep and cattle;
- the development of the woolly undercoat, especially in sheep, and hairlessness in the pig;
- docility, and the ability to cope with crowding stress;
- earlier pubertal development, an increase in the length of the breeding season and an increase in fecundity; and
- in farmed salmon, adult body size, egg size and time spent at sea have changed with selection, especially for growth rate (Teletchea and Fontaine, 2014).

Compared with wild animals, farm animals are managed and consequently (see Fisher and Bryant, 1993):

- age, sex, social and kin structure are distorted or disrupted by culling, grouping animals by age or sex, and by management practices such as castration;
- movement and migration may be limited or prevented by fences and management practices;
- animals are less subject to predation but subject to systematic culling and harvesting;
- exposure to ill-health and disease varies but animals may receive veterinary care;
- food and nutrients are often less varied in composition and seasonal quality but may be supplemented;
- aggression between individuals may be varied by altered group composition and resource availability;
- reproduction and sexual selection is controlled by reduced choice of mate and manipulated timing of the mating season;
- parental care of young may be supplemented by humans, or curtailed by weaning on separation of the mother and her young; allo-parental care may be disrupted by altering kinships or increased stock densities;
- there is selection for more tractable (tame) and bigger, faster-maturing, more fertile and fecund animals along with better nutrition and health and husbandry;
- there are reduced opportunities for resources in relatively more impoverished environments, e.g. shade and shelter, but these may be supplemented; and
- there is increased contact with and intervention by humans and other species (e.g. dogs and horses) associated with man.

The genetic, physiological, behavioural and morphological changes are all biological; the possibility of domestication affecting animal culture, that which is learned and transmitted, is relatively unknown (Clutton-Brock, 1992). For instance, hefting involves keeping animals on land or pasture to which they are accustomed. Another example is that of African bushmen's cattle: when a foreign bull was introduced it disrupted herd behaviour and some were killed by lions. Previously, they had walked out to graze in single file, varied their direction and returned well before sunset.

While biological changes in animals, and evidence of practices such as selective slaughter of young animals, are symptomatic of human management, a more complex and meaningful picture of domestication is revealed considering the social and economic forces shaping man and animals. Maintaining human communities contributed to more intensification of animals, in turn creating resources for humans.

The Neolithic Revolution was accompanied by a ten-fold increase in the human population, displacing or assimilating the less dense population of hunter-gatherers. Plant and animal domestication led to more food that, when stored, resulted in denser populations. Along with the use of animals for labour and transport, these events reinforced and facilitated settled, politically centralized, socially stratified, economically complex, technologically innovative societies (Diamond, 1997). Farming also had costs, what Diamond (1987) has called a double-edged sword. The change from the varied diet provided by hunting and gathering to a more limited diet provided by farming, brought a decrease in average height and an increase in anaemia and dental decay. There was also a greater risk of starvation associated with dependence on fewer crops. And the evolution of diseases and parasites in dense populations of malnourished, sedentary people, re-infected by each other and their own sewage, made many worse off (Diamond, 1991; Tudge, 1998). Farming was also hard work: the more land, plants and animals are manipulated the more food is produced. In contrast, if a hunter works harder, the prey available decreases. Thus, predators and hunters tend to be lazy; lions hunt for perhaps two hours a day, Kalahari bushmen about six hours a week. There was no turning back from farming. Increasing numbers of people and declining fauna meant that people had to rely increasingly on farming for their food. The more people farmed, the more the population increased and consequently the more people needed to farm since it was the only way of feeding the extra population. 'To condemn all of humankind to a life of full-time farming... was a curse indeed' (Tudge, 1998). The modern preoccupation with having to milk cows twice a day, every day, certainly adds weight to that observation.

Slash and burn methods cleared forests to make land fertile for cultivation in many parts of the world, and in the Nile Valley floodwaters and irrigation were used to maintain fertility. In other areas, open grassy and bushy savannas, steppes, prairies, tundra and highlands were more immediately suitable for pastoral societies. Cattle were bred in the Sahara, yak in central Asia, goat and sheep in the Mediterranean and Near East, llama and alpaca in the Andes, and reindeer at higher latitudes. Although both cultivation (farming) and animal breeding (pastoralism) overlapped, their respective societies began to differentiate. For some, livestock, especially cattle, became a means of accumulating wealth and power, and of displaying social relationships built on obligations and debt. The accumulation of cattle for bridewealth and for strengthening political relationships is a characteristic shared with some modern cattle-keeping cultures (Mazoyer and Roudart, 2006). Self-aggrandizement, the acquisition of property and the gaining of social advantage or power were important drivers of farming, evident, for example, in feasting, the importance of dogs, and bottle gourds that were not sources of food but important for the accumulation of wealth, display and power.

Increased food production and storage allowed for the development of craftsmen and artists, soldiers and scribes. And, more ominously, the change from hunter-gatherer to farmer brought the opportunity for a 'class of social parasites who grow fat on food seized from others' (Diamond, 1991). Stockpiled food enabled the elite to gain control of food produced, assert a right to tax others, escape the need to 'feed' itself and engage in political activity, bureaucracy and the production of luxury goods. Food became a commodity and those responsible for its production and distribution gained wealth and power. Whether farming was the instigator of these changes is unknown. Adopting farming does not necessarily require social divisions such as the subordination of women or the keeping of slaves, any more than does the Industrial Revolution. It may be that it provided an opportunity for those involved to develop or utilize them. Farming, both cultivation and pastoralism, worked because it produced more food from the environment than would otherwise be the case, allowing the human population to survive and increase. Domestication increased energy surpluses and an agricultural economy supporting the development of urban life and a growing culture. It also provided benefits and favoured people who worked hard whereas hunters (human and animal) are not similarly indefinitely rewarded for their efforts. Large populations of hard-working people inevitably prevailed, hard work associated with an exponential rise in the human population.

The domestication of animals, an evolving relationship and investment by both partners, was part of this success. Understanding the full richness and diversity of the partnership may help understand the place of animals in modern society. It has, as evident in the recent domesticate, red deer, important environmental, demographic and social components and is more than just taming and changing animals for human benefit. The road to modern farming was, as Barker (2006) notes in *The Agricultural Revolution in Prehistory*, about a combination of psychological, technological, social and economic changes in the way humans thought and did things. The hunter-gatherer sharing, or egalitarian way of life, was replaced by a more egotistical ethos reflecting a work ethic. Animals and humans coevolved, the relationship domestication. It was an evolutionary strategy, a life better than life in the wild.

The agricultural revolution of the Middle Ages

How they toiled and sweated to get the hay in! But their efforts were rewarded, for the harvest was an even bigger success than they had hoped.... The pigs did not actually work, but directed and supervised the others.

After its Neolithic beginnings, the Mediterranean and European ecosystem developed around four settings: woodlands, common grazing areas, fields of cereals, and gardens and orchards. It was a mixed system with livestock grazing fallow land or penned there nightly to ensure the land was fertilized with dung for future crops. To prevent them grazing cereals, herds were guarded, tethered to a stake in the ground or kept in enclosures made of bushes or wooden hurdles. In addition, animals, especially cattle, donkeys and mules, were used to draw an ard, a hoe-like light plough creating a furrow for planting seeds. The relatively plentiful labour meant food production could be enhanced by drainage, terracing, hoeing and tilling.

For thousands of years from the Middle Ages until relatively modern times, much of England and Europe was occupied and cultivated by openfield systems. Poultry, sheep, pigs and cattle were kept, and barley, wheat, rye, peas and beans grown, with woodlands providing acorns and beechmast for pigs and some fuel. The Nottinghamshire village of Laxton (Figure 2.4) is a remnant of that system. The village community (the Domesday Book suggests 35 adult males out of 100–120 people) farmed the land around them, providing for their own subsistence through a common operation. The land was ploughed in long strips, each man's strip alternating with his neighbours', sharing the best and worst land. Cropping provided corn for humans and straw for livestock with woodland and fallow land providing grazing. Livestock similarly fitted in, grazing the commons and the aftermath of the meadows in summer and autumn, supplemented by stubble and meadow after the harvest, hay and straw in winter, and

Figure 2.4 Parts of the Laxton map, a representation of medieval open-field farming in 1635 (*The Bodleian Libraries, The University of Oxford*; MS. C17:48 (9) Sheet 7, south-west corner detail). The village of Laxton north-east of Nottingham is the only one in the UK still practising open-field farming. The 450 acres of farmland are divided into three large unenclosed fields farmed in strips on a three-year cycle, one lying fallow each year.

the grass on the commons in spring. The 108 acres of commons at Laxton provided highly valued and jealously guarded individual grazing rights for 20 sheep (or an equivalent in beasts and horse) until 1908 when it was reduced to 10 sheep, 3 beasts or 2 horses. Livestock were kept on commons enclosed with hurdles, or tethered. Sheep had to be washed before shearing to clean their fleeces and it was hard physical work before the advent of horse-drawn harvesting. The system was governed by Manorial law that oversaw the farming year, fines being incurred when, for example, animals were turned out too soon, or if someone grazed more than their share. The pinder, responsible for rounding up stray animals, and the forman, responsible for overseeing farm activities, were two of the most important roles in the village (Beckett, 1989; Dewey, 1989).

Open-field farming was finally undermined by the need to respond to rising demand for the produce of the land and to do so by agricultural specialisation. Shared land was enclosed, becoming private property. This enabled farmers to experiment with livestock breeding and stocking densities, as well as pastures and fodder crops. The improved methods raised production, reducing food prices and feeding the growing population. In

England, by 1850 only 20% of the population were directly involved in farming, producing food for themselves and the remainder of the population. Beginning in the 11th to the 13th centuries, and especially in the 16th to the 19th centuries, farming was transformed by many technologies and practices such as the following:

- the use of the plough, which vastly improved soil fertility, the wheeled cart and harvesting machinery;
- cultivation and animal breeding were integrated, grasses and fodder crops replacing fallow periods;
- selective breeding of livestock for production, and labour and technology for enhancing ploughing, e.g. the horse collar and iron horseshoes;
- housing (cowsheds, stables, sheepfolds) enabled animals, particularly cattle, to be wintered in stalls and their manure to be transferred to the fields;
- conservation of fodder enabling winter carrying capacities to be increased, contributing to a general doubling of livestock numbers; and
- mechanization, e.g. the replacement of animals with steam engines, eventually further revolutionized labour, farm production and transport, bringing fertilizers and connecting farming with distant markets.

Life was very busy involving haymaking, cleaning out of stabling and putting in clean litter, feeding and watering livestock, guarding grazing animals and leading pigs into the forest to fatten on acorns and beechnuts. There was lots of other work too: harvesting grapes and making wine, pruning vines and fruit trees, repairing fences, cleaning out ditches and streams, gathering firewood and timber, gardening, gathering, hunting and poaching, repairing roofs and buildings, spinning and weaving, grinding grain, baking bread, salting and smoking meats and so forth. But there was also the need to care for twice the number of livestock, cart and spread more manure, and cut, bundle, transport and thresh larger harvests. There were also setbacks, enclosures and famines, the Black Death killed large numbers and contributed to the loss of communities – 3000 villages in England alone (Pretty, 2007) – and mixed farmlands were converted to sheepwalks. The relative abundance of consumer goods meant there was little motivation to work for extra income and people only worked long enough to meet their needs. While the population was once kept in check by food production, this restraint lessened at the end of the 18th century, the population increasing and food prices falling. Europe's agriculture was

Figure 2.5 Sheep bred by Robert Bakewell as portrayed in John Ferneley's 1823 painting of 'Sir John Palmer on his favourite mare with his shepherd, John Green, and his prize Leicester Longwool sheep' (*Courtesy Leicestershire County Council Museums Service*). To satisfy the demands of the growing human population of the Industrial Revolution, Bakewell bred fat animals that matured earlier. (Available in colour in the plate section between pages 178–179.)

further disrupted by its colonies having lower production costs and, along with cheaper transport, this resulted in the importation of large marketable surpluses.

While livestock selection is as old as farming itself, it was Robert Bakewell (1725–1795), an 18th-century Leicestershire farmer, who publicized the idea of selecting stock for economic performance (Pawson, 1957; Wykes, 2004). Perhaps the most well-known of England's breeders, Bakewell focussed on breeding sheep (Figure 2.5) and cattle, as well as horses and pigs, for the marketplace. What butchers wanted was meat and fat: 'I do not breed mutton for gentlemen, but for the public.' Previously, breeders had emphasized traits such as head shape, horns and colour.

The New Dishley or New Leicester's major trait was early maturity, yielding the most meat in the shortest time and at the least cost. Bakewell's animals were finished in 27 months compared with the 39 months of other breeds of the time. He achieved this by separating the sexes to prevent indiscriminate mating, inbreeding to fix and exaggerate desirable traits, measuring feed consumption and carcass properties, rigorous culling, and observing how well sires did in different environments. He let rams at high prices to ensure his sires were mated to the best. Over this period, there was a striking improvement in the weight of animals butchered. For example, sheep weights at London's Smithfield Market (Pawson, 1957) increased from 28 to 80 lbs between 1720 and 1795. Sadly, Bakewell was bankrupted by social extravagance, financial mismanagement, the costs of experimentation, and his love of hospitality and travelling to learn about farming. His breeds have largely disappeared, the fate of a failure to preserve traits such as milk yield and fecundity, ability to handle poor weather, and to maintain their mobility.

Many factors contributed to the massive increase in food production, by two and a half to three times in England between 1700 and 1850. Land, for example, wetlands and woodland, was claimed and improved; more and higher-yielding crops were grown, both for humans and for grazing animals (instead of leaving land to lie fallow); arable, pastoral, dairying and livestock fattening became specialized in those regions best suited to them; and the plough, seed drill, scythe and threshing machine also brought improvements as did changes in labour practices (fewer people were employed). Finally, there was an improvement in the ability of farmers to farm with the least successful leaving the industry. It was also accompanied by significant changes in social life.

The agricultural revolution also brought dramatic changes in landholding and social relationships. Enclosure, the disappearance of common property rights, reflected the replacement of a predominately subsistence-orientated rural economy with one firmly linked to the market. The consequences of enclosure for people were captured by Thomas More (1516) in *Utopia*. Sheep once 'naturally mild and easily kept in order, may be said now to devour men, and unpeople, not only villages but towns'. The promise of 'softer and richer wool' resulted in many people being dispossessed of the land for grazing. These changes affected the fortunes of various social groups in the countryside, particularly in their relationship to the land. They included a rise in landholding by the gentry and the great estate owners, a decline in the proportion of land owned by yeoman freeholders or small farmers, and a rise in the landless class of farm workers, the agricultural proletariat. Between 1455 and 1637, 34,000 families were dispossessed, some becoming beggars, others migrating to cities, emigrating or becoming industrial labourers. Labourers started becoming worse off, women's work declined and the market system made it easier to transport produce abroad for more lucrative returns.

Sixteenth-century farming was dominated by relationships of kinship and neighbourliness (people saw each other as equals), and paternalism and deference (unequal obligations). The latter relationships were based on an acceptance that society contained permanent inequalities, therefore those with wealth and power were expected to help those less fortunate. In return, deference and obedience usually served to maintain stability in rural societies (Overton, 1996). However, reflecting the imperatives of the market economy, poverty became the responsibility of the individual and there were outbreaks of violence and unrest associated with food availability. Agricultural labourers demanded a 'moral economy' to respect the rights of the poor who were increasingly dependent on the market in the face of declining subsistence farming. Rural unrest was epitomized in England by the Captain Swing riots of 1830 (Hobsbawm and Rude, 1969). It was an era as much about trade unions and power as about the replacement of men by machines (Thomis, 1970). Prior to industrialization, farm labourers' earnings were comparable to others. However, by the end of the 19th century they were among the lowest in society as the demand for non-agricultural products rose more rapidly than for farm products (Dewey, 1989). By 1914, 60% of the UK's food was imported. At a time when labour was plentiful in England, and harvests poor, hungry labourers, inspired by the mythical Swing, resorted to breaking threshing machines, arson, robbery, rioting and making threats in demanding a right to employment and a minimum living wage, as well as a reduction in rents, tithes and taxes. Consequently, some 1400 were sentenced to jail, death or transportation to Australia. A reaction to modernization, both economic and social, it was also a response to agrarian distress and the need for constitutional reform borne of over-taxation, want of work and insufficient wages (Wells, 1997). The tension between farm labourers and owners is now ingrained in the labour movement, with such stories as the Tolpuddle Martyrs (Jones, 2000), and such images as the *Skeleton at the Plough* (Price, 1874), part of England's social history.

Middlemen and commerce

Every Monday Mr Whymper visited the farm.... [He] was ... sharp enough to have realized earlier than anyone else that Animal Farm would need a broker and that the commissions would be worth having.

Nottingham's annual autumn Goose Fair is one of the UK's largest and most well-known funfairs. Held in October, its origins were probably in a holy day that, because of the opportunities for trade and commerce, developed into a market in the Danish settlement more than 1000 years ago. Although it became famous as a cheese fair, in the Middle Ages it was customary to eat goose at the Festival of Michaelmas, a celebration of the end of the harvest and the first day of the new farming year. A last fling before winter, it was also an opportunity to seek winter employment. At that time of year, geese were plentiful and in good condition, having hatched in the spring. Up to 20,000 were thought to have been walked to the fair from neighbouring counties, the birds' feet protected with a mixture of sand and tar (a modern re-enactment of the trip with 20 birds saw them arrive in better condition than when they started). Eating goose was believed to protect families financially for the following year, Michaelmas Day being one of the four days of the year when rents were due. While the original purpose of the fair may have been trade in geese, cheese, pigs and turkeys, the character of the fair has changed. Nineteenth-century improvements in transport, distribution and retailing meant it was less important to stock up with goods for winter. Gradually, the Goose Fair largely became a pleasure festival with the introduction of exhibitions, entertainments, amusements and bizarre sideshows. Fortune-tellers, jugglers, acrobats, wild beast exhibitions, freak shows, merry-go-rounds and travelling menageries have given way to roller coasters, dodgems, waltzers and the newer rides of today. Goose and cheese have been replaced by takeaway foods though the candy Cock-on-a-stick, possibly originating from a Goose-on-a-stick, remains (Lund, 2005).

The evolving nature of the Goose Fair reflects societal changes of the agricultural revolution. Not only the development of formal markets for exchange of produce, but the rise of the middlemen: badgers, kidders, jobbers, drovers, shipping merchants, brokers, auctioneers, peddlers and other indispensable roles. Livestock middlemen (Figure 2.6) included graziers who fattened animals on grass, turnips, oil-cake or other products. The jobber bought and sold, the drover dealt in cattle and driving them to the market, carcass butchers served wholesale butchers and meat dealers and sold to retail or cutting butchers. Cutting butchers bought personally or from carcass butchers, while hucksters cried out their meats (rather than operating a stall). Poulters dealt in rabbit, game, eggs and birds. Middlemen became specialized, processing food, for example, millers and bakers, dealing in grain, for example, corn merchants and cheesemongers, and giving credit at a time when banking was basic.

Figure 2.6 An example of livestock and meat middlemen in England during the 17th and 18th centuries (redrawn from Westerfield, 1915).

Prior to the growth of middlemen, the ideal of the market was one in which farmers traded their excess food directly with consumers; surpluses were distributed to those who needed them. Middlemen saw and created new opportunities, locally and internationally, took the initiative in organizing, correlating and establishing a mutual dependence between producers and consumers, and influenced the policies of the state to the furtherance of commerce. This active, thinking, organizing, constructive class was instrumental in, if not fundamental to, the development of growth and commerce. The successive differentiations of middlemen, each specializing in the performance of more limited functions, have contributed much to the progress of society. However, at least initially, there was much hostility to the rise of 'middlemen' who bought to sell again, disturbing the ideal of distributing food surpluses to those in need. They were thought to keep prices high and cause food shortages, and not surprisingly were generally reviled, regulations restricting their activities.

Where most people had once produced their own food with any surpluses distributed locally via a medieval market, a competitive system began to arise where produce was distributed in time and space depending on what consumers could pay, as societies became less dependent on climatic variations disrupting farming harvests. Economic life acquired a speculative dimension with farmers now separated both in distance and in lifestyle from consumers. Aided by the farming of livestock better suited to transport to distant markets, for example, geese, and by road, rail and water transport networks, large and distant population centres were supplied, which, in turn, added to their growth and an increase in the number of people not directly working in agriculture. For instance, NZ's agricultural economy acknowledges the first shipment of frozen meat, mostly sheep carcasses, that left on the SS *Dunedin* bound for London as something that shaped the country more than any other event. The voyage was not without drama. Concern for the adequacy of the circulation of cold air necessitated cutting new air duct openings whilst the ship maintained a long tack, and saw the Captain nearly freeze and having to be pulled out by a rope attached to his legs. The refrigeration equipment burned three tonnes of coal a day and sparks from the funnel threatened to set the sails alight several times. However, the consignment arrived in London in February 1882, at a time when meat supplies were low, and the carcasses 'as clean and bright as freshly killed mutton'. They sold within a fortnight at twice the price they would have made in NZ (Williscroft, 2007). It began a business that contributes hugely to the country's economy and sees NZ as the world's largest exporter of sheep meat, and the second-largest exporter of beef.

Gradually regulations restricting middlemen broke down as market activity increased and subsistence farming declined. With time, the number of middlemen increased, and they became more efficient and specialized, giving birth to business and thus profoundly influencing social history. Poor communication, at least by modern standards, meant middlemen knew more of supply and demand and could trade much more effectively than could farmers or consumers, and they could distribute surpluses that would otherwise not be able to be fully utilized. Middlemen, then, essentially had two functions: one the geographical connection of producers and consumers; the other connecting farmers and consumers in time, a function requiring capital and credit enabling the supply of goods to fit demand. Matching supply and demand required foresight, storage facilities, an organized market and business capital. As the role developed, value was added to products, and bigger dealers came to dominate as a capitalistic class. Meanwhile, farming and farmers changed from being largely husbandmen supporting the family (i.e. mostly subsistence production for their own needs, but with some dealings with the local markets) to entrepreneurs largely supplying larger, urban and wealthier markets. Money was now being made by respectable citizens through buying and selling where it had once been restricted to hucksters or low-class persons (Polyani et al., 1957). The market began to become a trade and middlemen rose in social standing, becoming part of the elite of society.

As human population growth accelerated in the 18th and 19th centuries, so did the demand for food. Market economies replaced subsistence agriculture as industrialized societies became more urbanized and less close to the land. While animals once played a major role in determining the quality of life of people, they now were a commodity to people removed from farming, consumers influencing animals' quality of life through their purchases. This was a marked and profound change in humankind's relationship with animals.

Modern farming

Snowball had made a close study of some back numbers of the Farmer and Stockbreeder which he had found in the farmhouse, and was full of plans for innovations and improvements. He talked learnedly about field-drains, silage, and basic slag ...

Humans have been improving agriculture throughout history but the most rapid or marked periods of intensification belong to the 20th century when 'farming left the farmyard to become an industry' (Appleby, 1999). The technical and infrastructural changes after World War II were particularly marked. For instance, the average British farm size doubled to 51.6

hectares in the period to 1981, and the number of farmers halved from the 438,000 there were in 1945. As non-farm employment became more attractive, easier and more accessible, farmers sought economies of scale to finance machinery, fertilizer, drainage, buildings and land (Dewey, 1989). Unlike the low stock density, outdoor farming systems of the last 10,000 years, large numbers of animals are now permanently housed in confined and barren environments, 'farmed' using a high degree of mechanization (e.g. feeding, watering, manure removal, climate control, milk and egg collection). In addition, standardized feed, genetic selection and enhanced husbandry contribute to more marketable, consistent and efficient production of meat, milk and eggs. There are now fewer farmers farming greater numbers of more productive livestock with greater inputs and technology. Following the industrialization principles of Henry Ford, farming has increasingly specialized. Mixed or diverse systems, while better at managing risks (e.g. a good year for grapes may mean a poorer year for livestock, so if you farm both you do alright), do not produce the economic yields of more specialized farms over time. Strange (1988), in *Family Farming: A New Economic Vision*, summarizes modern intensive farming in the US as:

- industrially organized, land being an investment expected to bring financial returns, and requiring capital for technology;
- large scale, concentrated and specialized;
- reliant on a greater proportion of inputs, e.g. fertilizers, feed, finance, management, coming from off the farm, and standardized production practices tending to remove or minimize ecological constraints normally associated with farming;
- able to seek advantages borne of specialization, the scale of the operation and the volume of product; and
- largely driven by the business values of economic efficiency, productivity, competition, progress and modernization.

The modern industrial focus is seen by many as contrasting with smaller farms providing for the livelihoods of families and their neighbours and communities.

Three features are particularly worth noting. The first is the changes on the farm. For example, mechanization enabled draught animals to be replaced by tractors and specialized machinery, freeing farms from having to grow fodder for horses and oxen but requiring them to source machinery and fuel. Other inputs such as pesticides and fertilizers freed farms from the necessity to rotate systems and meant stocking densities could be increased. Furthermore, it meant strains and breeds of animals could be selected that were best adapted to the production system and farm profitability. For example, cows best adapted to machine and robotic milking tend to be retained in herds. Similarly, in selecting animals suited to the ecological constraints of the system, good husbandry, and the requirement to remain profitable, there are the demands of retailers and consumers, for example, for fat or lean meat. The second feature is the remarkable reliance on a multitude of industries both upstream and downstream of the farm: industries that produce fertilizers, pesticides and machinery, and those that process, market and sell animal products, as well as, for example, commerce, transportation, consulting and assuring. These service industries or agribusinesses probably account for two or three times as many people as are directly involved in farming. Like farming, they too have become larger and specialized. While the transfer to agribusiness of many tasks has relieved farmers of them, it has meant that the farming system is complex, and not just that defined by the farm boundaries. When we think of farmers, should we also have to think of the service industries that farming is wedded to? The final characteristic is that it is farmers themselves who must choose and combine the various inputs and develop the most advantageous production systems dependent on the constraints of their land, climate and animals, their financial and personal circumstances, and the threats and promises of markets and society. As Chapter 4 will explore, there is no shortage of advice for, or demands of, how farmers should farm. Nevertheless, the paths that some farming systems have taken, especially those confining poultry and pigs in barren environments, are the subject of some contention and it is worth noting the nature of those paths.

Although chickens were initially important as sacrificial animals and in cockfighting, and the Romans farmed them, they were considered an alternative, not the stuff of mainstream agriculture that focussed on cereals and meats. For instance, in the US chicken farming was the domain of women and slaves. Usually kept in small flocks outdoors on mixed farms, it was not until the 19th century that large-scale husbandry became popular (Wood-Gush, 1959) leading to the industrialization of poultry farming. The production and hatching of chicks, manufacture of feedstuffs, the raising, slaughtering and processing of birds, and the marketing of chicken meat by vertically integrated companies, arguably have their roots in industrial management systems, especially the steel and automobile industries. No doubt there were many entrepreneurs and stories, but those of two individuals are worth recounting. Growing vegetables for sale in the Great Depression, John Tyson was often paid in chickens that he marketed in towns. As his market expanded, he provided the birds on his vehicle with feed troughs and drinkers so that they maintained their condition on longer trips. Eventually, after buying, transporting and selling, he moved to raise and process birds on his own account (Morrison, 1998). Another,

Figure 2.7 Celia Steele, two of her children and broiler caretaker Ike Long, with chickens and their houses during the early days of the commercial broiler industry in the US *(Photo courtesy Delmarva Poultry Industry, Inc.)*.

Celia Steele (Figure 2.7), in 1923 received 500 chicks from a hatchery instead of the 50 she had ordered. With her husband, Mrs Steele raised the birds as broilers or meat chickens inside a small wooden building heated by a coal stove, the first commercial broiler house in the US. Success led to the raising of many more chickens for their meat and, with breeding to find the perfect carcass, eventually the modern industry supplying much of the world's protein.

With increasing intensification and specialization, flock size increased and with the birds kept indoors but with access to outside. Parasites and diseases were common problems since many birds were kept on a limited area of litter. While floors were developed to separate hens from their faeces and minimize the risk of disease, there was significant risk of hysteria, feather pecking and cannibalism. The first cages housed single birds in the 1930s, and though there were early injuries when birds became trapped, hygiene improved, diseases virtually disappeared, and feather pecking and cannibalism were reduced. While many large flocks were kept in deep-litter and wire-floor houses in the 1940s, the following decade saw a rapid and sustained move to cages for easier management. Multi-bird and tiered cages became popular enabling a higher stocking density. The benefits of higher house temperatures, controlled lighting, along with insulation and ventilation, led to automated controlled environment housing along with mechanized feeding, watering and egg removal. Changing attitudes to the cages, and the willingness of some consumers to seek out and pay a premium for non-cage eggs have seen the poultry industry invest in alternatives to battery cages. Modern poultry farming is an outstanding success story; a virtually new industry was created after World War II. With an unspecialized flock almost universally found on farms and a market saturated for eggs, farmers turned to cost reduction and the battery cage system. The breeding sector turned to mass production, and the meat sector to compound feeds and the breeding of fast-growing birds.

As with poultry, the modern breeding and confinement of pigs have advantages (Thu and Durrenberger, 1998). Housed animals are protected from the weather, especially when heat is provided in winter and cooling in summer. Feed efficiency and weight gain have increased, animals are protected from some diseases, injuries and ill-health, and feeding, watering and manure removal are automated. In addition, animals can be bred all year. However, there are also disadvantages: not only the impacts of confinement on welfare, and the need to manage greater amounts of manure, but the higher stocking density risks the rapid spread of diseases necessitating increased vaccination and antibiotic use. The capital investment requires increased herd efficiency and/or productivity to meet financial obligations, and the increased labour requirements associated with, for example, continual farrowing mean a greater reliance on hired employees. Thus, modern pig farming is also complex; farms are specialized, larger and fewer, more intensive, mechanized and involving fewer people. However, there are concerns around the health of farmers and their families, and the environmental and social impacts of the industry in some regions.

The developments in modern pig farming in the US – the large, specialized and corporate farming systems – are the more recent examples of wealth being drawn from agriculture. A profit-driven, industrial-like paradigm has been so successful for some parties that it has raised the question, 'Why own a farm when you can own the farmer?' The contributors to Thu and Durrenberger's (1998) *Pigs, Profits, and Rural Communities* lament the power of meat processors and the different price structures favouring larger farms, and lower profits, underlying drivers of animal confinement and automation to reduce costs. They also lament the consequences of this form of farming for rural communities, especially their social, economic, human, environmental and political well-being. Modern transport and economies of scale are among the features of western economic systems that have enabled businesses to procure animals from a wide area, process

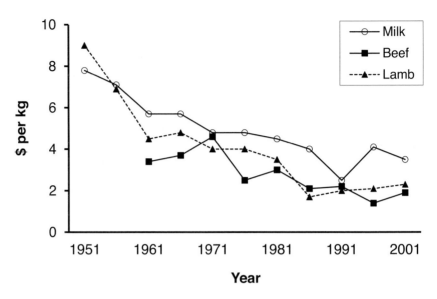

Figure 2.8 An example of the decline in farm-gate returns in NZ over the second half of the 20th century. Similar patterns are evident in egg, pig and cattle in the US (Norwood and Lusk, 2011; Plain, 2010).

them and to sell the products widely. As more food is produced, farm-gate returns decline as the price paid by wholesalers, retailers and consumers falls (Figure 2.8).

Increased intensification is essential for maintaining business viability, underlying the specialization (and a corresponding loss of the advantages of diversification) and capital investment characteristic of systems such as modern pig and poultry farming. It is not only the returns for produce that have driven the intensification, but costs have also risen. For example, on one mixed farm in England, the income from 2.1 lambs was required to pay the annual accountant's fee in 1958, but in 2006, 40 lambs were required. Similarly, the cost of a vet visiting and treating three cows was equivalent to the income from 36.4 gallons of milk in 1958 but 176.5 gallons in 2006 (Cooper, 2006). This is not just limited to animal farming – a tractor cost the equivalent of 21 tons of sugar to a Jamaican cane farmer in 1972, rising to 50 tons in the 1980s (Body, 1991). Paying for such increases requires larger, more productive and efficient farms, the 'wealth' of farming being 'shared' with the wider agribusiness community, as well as consumers. It has been likened to a vicious cycle or treadmill. The lure of technology resulting in new systems designed to make farming more efficient and profitable, generally requires larger farms and more productive animals to pay for that technology. And as tasks become more automated, there is less job satisfaction and a greater reliance on immigrant labour. Similarly, while early adopters of technology gain advantages, others must eventually follow suit just to remain competitive, the manufacturers of the technologies benefiting, along with consumers and agribusinesses. While it can make good economic sense to use science and technology, it is not always benign in helping to shape farming and may bring both benefits and problems, for example, antibiotics bring relief from disease and the risk of resistance.

A great surplus and a modern elite

There, round the long table, sat half a dozen farmers and half a dozen of the more eminent pigs, Napoleon himself occupying the seat of honour at the head of the table.... The creatures outside looked from pig to man, and from man to pig, and from pig to man again; but already it was impossible to say which was which.

Land use has changed through history from being a source of animals for hunting, to a place to use labour to produce food, to a place where capital and technology can be used to generate money and power. As humans settled and then began farming they separated themselves from the wild and continued on the path to civilization. They needed to defend stored foods and coordinate activities, no longer relying on egalitarian exchanges but establishing rules of ownership, caste structures and deities to make it possible to deal with strangers (Clark, 1999). Agriculture was not just the growing of food, but the accumulation of wealth, power and status; owning and working the land were held to be of social importance.

The need to coordinate activities in growing, storing and trading food surpluses resulted in a new social and moral order. Tensions began to develop though. While hunter-gatherers may have viewed themselves as part of nature, farmers acted to overcome and subdue nature's 'wildness' and 'savagery' (and saw themselves as separate from the natural world). The second half of the 20th century saw an acceleration of the patterns described for the first 10,000 or so years of farming. Developments brought benefits in food production, culture and civilization, and changes in social worldviews and beliefs. Nowadays, we derive much of our wealth from agriculture yet few of us are directly involved in farming, and even fewer are capable of being self-sufficient. The farming sector, the original primary producers, have been joined by the processing, wholesaling, retailing and service sectors.

The change from growing produce for one's own use to growing it for economic gain is now nearly complete; farming also serves to make money.

The business of farming, the drawing of wealth from agriculture, has now become inseparable from the production of food to feed humans. The new middlemen add value, and more recently add higher added value, to farming's products. Many affluent and western farmers now obtain much of their own food from the marketplace. Purchased pasteurized milk is preferred to milking a house-cow and few slaughter their own animals for meat, preferring the convenience afforded by the butcher. There have even been moves to prevent farmers slaughtering stock for their own use, or weeding their crops by hand, developments that effectively amount to a legal requirement to use middlemen.

Within an entirely agrarian society, consisting of four mouths to feed per worker (including the latter), productivity must be at least equal to four times the dietary needs of the average individual. This is because those supported by the worker, in any society, include children and the elderly, those not producing their own food. A non-agrarian society composed of as many farm workers as non-farm workers requires each farm worker to feed eight individuals. Surpluses produced beyond that required to support farm workers and their families, provide the capacity to support dedicated warriors, priests, administrators, artisans, merchants and so forth. (Mazoyer and Roudart, 2006). Through history, the amount of food produced by those on the land has increased, supporting others. For example, in 1820 each person on a farm in the US produced enough to support him or herself plus 0.25 of another person, by 1920 it was 2.5 others, and in 1945, each person supported 5.5 other people (Cooper et al., 1947). The increasing surplus of production, facilitated by the adoption of mechanization especially, provided for the increasing necessities of life: food, clothing and shelter. A continued and stable surplus, along with the capacity to process raw materials, trade and commerce, some form of stable government, customs and religion and leadership, is required for civilization to keep advancing. Most are conditional factors, but the absolute requirement is that primary producers must produce a surplus. The factors that determine the amount of this surplus largely determine a civilization's status.

There are now few people directly engaged in farming. For instance, in the UK in 2007 there were 526,000 farmers and farmworkers, less than 1% of the total population. In contrast in 1870, half the US population was employed in agriculture. However, there are many more involved in processing and delivering food; globally some 1.4 billion farmers help feed 7 billion people. For example, of nearly 3.9 million people involved in providing the 63 million people in the UK with food, only 480,000 or 12.4% were farmers, the majority involved in processing (10.3%), wholesaling (5.2%), retailing (30.9%) and serving (41.2%) food (Lang and Barling, 2013). The extent of those involved in processing and marketing food is

Table 2.2 *The distribution of the costs of woollen products, manufactured from wool grown in NZ and exported to the US for tufted carpets, and the UK for fine knitwear, in the year 1996/97.*

Wool for carpets	$NZ per kg	Fine wool for knitwear	$NZ per kg
Farmer	3.79	Farmer	7.50
Exporter/scourer	1.03	Exporter/scourer/topmaker	3.68
Spinner/dryer	8.37	Spinner/dryer	12.11
Manufacturer	13.14	Manufacturer	24.26
Retailer	29.84	Retailer	43.54
		Tax	15.87

also evident in how the financial value is apportioned across the supply chain (Table 2.2).

Once hucksters, jobbers and kidders, the new middlemen are seed salesmen, butchers, bankers, scientists, information technologists, retailers, marketers and so forth. The value these middlemen add to farming varies. In lower-income countries commodity prices and consumer food prices are closely aligned. In affluent countries, the contrast is marked; in 2007, just 18% of the retail price of cereal products (bread, biscuits, breakfast cereals etc.) was accounted for by the price of wheat. In contrast, some 50% covered 'marketing' – packaging, advertising and retail margins (Ritson, 2008). In the US, in 2001, over 80% of the food dollar went on services and materials added after the farm gate, that is, transportation, processing, distribution, labour, packaging and energy (Capps and Park, 2003). It is no wonder that some have likened service industries living off farming to 'economic parasites', especially since 'the real profits of the last quarter-century have gone to the managerial organizations – the middlemen – wholesalers and large retail distributors of machinery, additives and bulk food' (Saul, 1992, 2005).

There are many examples of the modern elite, those living off the wealth of farming or civilization. The most entertaining account is probably the fate of the Middlemen, the useless third of the population of the planet Golgafrincham in Douglas Adams' *The Restaurant at the End of the Universe* (1980). The remaining two-thirds of the population, the Thinkers and Doers, tricked the worthless Middlemen, hairdressers, television producers, lawyers, telephone sanitizers and others, and ejected them from the planet. In modern times, the elite range from politicians, financial institutions and banks, to managers and celebrities, sportsmen and, in a way, all of us, farmers, teachers, nurses or whoever, who draw comfort from the essentials and luxuries modern civilization provides, from running hot water to holidaying around the world. In a sense, what we once knew as primary, manufacturing and service industries, have become 'primary' industries; instead of servicing farming, wealth is increasingly drawn from others, as a primary industry draws it from the soil and water, the sunshine, plants and animals. This is perhaps the most significant morally-relevant difference between people and animals: the latter are grounded in biology; the former, as we shall see, are increasingly virtually disconnected from that biology. When we lose the link, we risk culture becoming a law unto itself.

But we should not forget that farming produces our food. The price, the abundance and the choices many of us accept, almost unaware, are huge. We have gone from 80–90% of disposable income or labour going on food in subsistence farming, to around 10%, at least in the west – other nations spend 25–80% of their income on food (Hodges, 2005). This is also reflected in the falling prices of food, both collectively, evident in the international food price index (Roser and Ritchie, 2018), and in individual products, one of the most remarkable examples being the chicken. In the period after the end of World War II, a meat bird took 13 weeks to grow to about 2 kg in weight and cost the consumer the equivalent of about £40 today. Modern chickens take six weeks to grow to slaughter weight and cost less than £2 (Appleby, 1999). The deflated average retail price of chicken has fallen in that time (1960–2009) by 58%, beef by 27%, pork 31% and turkey 65%, and the returns to farmers for meat animals have more than halved (Plain, 2010). Put another way, we, the elite, now have real wealth and food is very affordable. In the Netherlands, you could buy seven eggs on the pay from an hour's work in the 1970s; in 2000, you could buy 400 eggs (Sandøe et al., 2003). Furthermore, many people can purchase almost any food at any time of the year, as the origin of fresh foods in a UK supermarket during winter shows (Table 2.3). In 1988, the UK produced two-thirds of its domestic food, but less than half in 2016, the contribution of imports from Africa increasing from 18% to 30% over the same period.

Characteristics of earlier parts of the first 10,000 or more years of farming are still apparent in modern times. The social unrest resulting in the Captain Swing riots has been replaced by London's *March for the Countryside*, which saw hundreds of thousands of people protesting mainly against opposition to hunting, but also highlighting farmers' falling incomes, advocating to 'Buy British food', 'Save our farms', 'Town and country not town over country' and the right to roam the countryside (e.g. Anonymous, 2002a,b). And in NZ, industrial disputes delaying meat plants slaughtering animals at a time when drought limited feed availability, saw 1300

Table 2.3 The origin of fresh foods in a single UK supermarket in December 2006.

UK

broccoli, cauliflowers, leeks, Brussel sprouts, parsnips, cabbages, mushrooms, swede, onion, potato, carrot, salad cress, beetroot, curly lettuce, shallots, turnips, pork, beef, chicken, fish, lamb, turkey, apples

Europe[1]	Africa[2]	Asia[3]	Americas[4]
fish, pork, beef, bacon; mushrooms, peppers, garlic; pears, apples, kiwifruit, tomatoes, avocado, raspberries, lemons, oranges; cauliflowers, cucumber, squash, courgette, celery; water cress, lettuce, Chinese leaf lettuce, little gem lettuce, spinach	baby sweet corn, beans, chillies, squash, peas; strawberries, grapes, plums, oranges, tomatoes	pomegranates, strawberries, grapefruit, okras, baby sweet corn, sweet potatoes, radishes, peppers, celery; coconuts; coriander, thyme, sage	apples, pears, avocado, blackberries, pineapples, bananas, cherries, blueberries, plums, mango, melons; sweet corn, asparagus, peas, ginger

[1] Spain, Norway, Germany, Ireland, France, Belgium, Italy, Greece, Poland, Holland; [2] Morocco, Canaries, Egypt, Kenya, South Africa; [3] India, Thailand, Sri Lanka, Israel; [4] Canada, US, Mexico, Costa Rica, Peru, Windward Islands (West Indies), Guatemala, Chile, Brazil, Argentina.

starving, 'old and worthless' sheep driven through a provincial town and killed (Anonymous, 1978).

While farming is the nexus of two worlds, that of biology, nature or the environment (e.g. soil, weather, animals), and that of people, communities and civilization (e.g. taxes, armies, markets), it is also arguably subordinate to both. We rely on farmers but we also, at times, dominate and exploit them (Tauger, 2011). Farming, in replacing hunting, gathering and foraging, became the interface between two worlds: the natural, from which we gain our physical sustenance and spirit; and the social or civilization, from which we gain much of what we deem important in everyday life. The relationship between farmers and the groups outside farming is crucially important to our understanding of animal welfare. The state of an animal, how it feels, reflects the world in which it is kept. That world is made up of a complex of historical, environmental, religious, economic, social and cultural components.

First civilizations were built on irrigated lands as they remained productive longer than those fed by rainfall, and were more dependable in catastrophic droughts (Dale and Carter, 1955). The developments at the farm level have been about retaining or enhancing soil fertility. Slash and burn farming has a limited span; when the soil is depleted farmers move on. A long period enables regrowth and the build-up of soil fertility. However, as the human population increases, there is pressure to reduce the interval between the successive slashing and burning of a plot. An alternative method of replenishing soil fertility is to use the silt-laden flood waters of major rivers, such as the Euphrates and Tigris in Mesopotamia, and develop sophisticated irrigation schemes. Another way is to maintain animals on the crop land during fallow periods, manure returning some fertility to the soil. Eventually, technology such as the ard and then the plough enabled this system to become more efficient, especially when manure from housed animals and phosphate from guano deposits were spread on the land. Finally, artificial manures or fertilizers were manufactured, freeing the dependence of the land on animals, enabling them to be confined on more efficient and specialized landless farms. As production increased beyond the needs of local populations, food, and the means of producing it, became a commercial or tradeable commodity susceptible to the whims of markets. As more value was added through the efforts of middlemen, and more food produced, farm income reduced (in a market, price generally reflects scarcity). Needing to maintain incomes, especially with the need to service significant debts, farmers' only options are to produce more by further intensifying their operation or getting bigger to achieve economies of scale.

Modern western society not only draws its food from agriculture, but also its wealth, or at least a significant part of it. Farming is not just about producing food, but also about distributing wealth. We have evolved from being relatively equal to specializing into different roles, some of which are more equal than others. As western civilization progresses, people move from agrarian to urban lifestyles and the proportion of the population farming decreases. With increasing urbanization, the agrarian population increasingly grows more food for the increasingly prosperous elite (Body, 1991). In the next chapter, the changes in the distribution of costs and benefits are examined, and how, with the ingenuity and application of science and technology (Roche et al., 2017), farming responds to these changes. Like Boxer, the draught-horse in *Animal Farm* who responded to the additional demands of the pigs with the personal motto 'I will work harder', the response has inevitably called for and resulted in harder and more complex work. In all of this are the pigs, the cows and calves, the sheep, chickens and deer. It is to their world, as well as the farmer, that we now turn.

Further reading

The Agricultural Revolution in Prehistory. Why Did Foragers Become Farmers? **Graeme Barker**. Oxford University Press, Oxford, UK, 2006.

The Agricultural Revolution. **J.V. Beckett**. Basil Blackwell, Oxford, UK, 1990.

A History of Laxton: England's Last Open-Field Village. **J.V. Beckett**. Basil Blackwell, Oxford, UK, 1989.

The Ascent of Man. **Jacob Bronowski**. BBC, London, UK, 1973.

The Way of the Animal Powers. Historical Atlas of World Mythology. **Joseph Campbell**. Times Books, London, UK, 1984.

The Deer Wars. The Story of Deer in New Zealand. **Graeme Caughley**. Heinemann, Auckland, NZ, 1983.

A Natural History of Domesticated Mammals. **Juliet Clutton-Brock**. Cambridge University Press, Cambridge, UK, 1999.

The 10,000 Year Explosion. How Civilisation Accelerated Human Evolution. **Gregory Cochran** and **Henry Harpending**. Basic Books, New York, NY, 2009.

Agriculture. **P. Dewey** In: Pope, P. (ed.) Atlas of British Social and Economic History since c. 1700. Routledge, London, UK, 1989, pp. 1–22.

The Rise and Fall of the Third Chimpanzee. **Jared Diamond**. Vintage, London, UK, 1991.

Guns, Germs and Steel. The Fates of Human Societies. **Jared Diamond**. Jonathan Cape, London, UK, 1997.

The Humans Who Went Extinct. Why Neanderthals Died Out and We Survived. **Clive Finlayson**. Oxford University Press, Oxford, UK, 2009.

A History of World Agriculture. From the Neolithic Age to the Current Crisis. **Marcel Mazoyer and Laurence Roudart**. Earthscan, Abingdon, UK, 2006.

The Prehistory of the Mind. A Search for the Origins of Art, Religion and Science. **Steven Mithen**. Phoenix, London, UK, 1996.

Agricultural Revolution in England. The Transformation of the Agrarian Economy 1500–1850. **Mark Overton**. Cambridge University Press, Cambridge, UK, 1996.

In the Company of Animals. A Study of Human–Animal Relationships. **James Serpell**. Cambridge University Press, Cambridge, UK, 1986.

Agriculture in World History. **Mark Tauger**. Routledge, London, UK, 2011.

Pigs, Profits, and Rural Communities. **Kendall Thu and Paul Durrenberger**. State University of New York Press, Albany, NY, 1998.

Neanderthals, Bandits and Farmers. How Agriculture Really Began. **Colin Tudge**. Yale University Press, New Haven, CT, 1998.

A History of Domesticated Animals. **Frederick Zeuner**. Hutchinson, London, UK, 1963.

CHAPTER THREE

High farming and hard work

The mid-19th century in England is regarded as a golden era of good-quality, excellent or high farming. It was mixed farming founded on the growing of fodder and cereals and the keeping of sheep and cattle, the latter often in yards or stalls. And it was intensive with high inputs and high productivity (Beckett, 1990; Jones, 1962; Overton, 1996; Perry, 1981). Intensification was achieved in part with the feeding of purchased oil-cakes to livestock, and artificial fertilizers to cropping land. Greater production also meant more animal manure and better crop yields. It was progressive, capital intensive and required new knowledge; progress was increasingly linked to soil fertility, new animal husbandry methods and social changes. High farming was not without its costs, though. For example, high productivity was achieved with mechanization at the expense of farm labourers, many of whom were malnourished and protested, epitomized in the Captain Swing riots (Chapter 2). Farming had moved from being relatively self-sufficient with farms producing their own grass, hay and straw to feed their own animals, who, in turn, produced enough manure to support the production of arable crops, to buying in feeds and fertilizer for 'processing' into saleable products. This was more akin to factory or industrial operations, producing for consumers in more distant markets aided by a growing segment of middlemen, dealers and food processors keen to innovate. Food production had become a business activity as much as a way of life, enabling farming to take advantage of market opportunities.

Although 'high farming' refers to agriculture in the 19th century, this chapter focuses on the high-input, high-output or intensification changes that have characterized farming, especially over the last 50 or more years: in today's language, 'top gear' farming (Body, 1984). It is most obvious in the change from mixed to specialized farming systems. For example, in the 1930s, a British smallholder over the course of a year might have picked sprouts, trapped rabbits, made kennels, netted salmon, mended boots, beat for shoots, sold pigs, cut osiers (willow shoots), driven at fairs, and sold mushrooms, plums, blackberries and wild ducks (Thirsk, 1997). At the same time, a 160-acre US family farm had milk cows, hogs and chickens and a large vegetable garden. It was labour intensive, neighbours working together to harvest crops, raise livestock and butcher animals for their families' meat (Nickles, 1998). Two decades later, a farm in the UK might have comprised 100 hectares, achieving most of its income from 30 or so dairy cows, and about 500 laying hens and 10 breeding sows, providing an affluent lifestyle for the farmer and his family (Webster, 1994).

However, within the last half-century, many farms have become larger, more specialized, more intensive, more capital intensive and less labour intensive (Donham and Thu, 1993). Intensification, along with concentration, regionalization, specialization, mechanization and new technologies, has resulted in fewer mixed farms, but larger farms, herds and flocks, fewer stockmen, more housed animals, greater stocking densities and higher-energy and/or higher-protein diets (Winter et al., 1998). For example, in 1992, Iowa (US) farms had five times as many pigs on over three times as large a farm as nearly 40 years earlier (Goldschmidt, 1998). Other changes in that time have included the replacement of purebred with synthetic genetic breeds; a focus on preventative health care for the herd rather than treatment of individuals; scientifically based management practices; computer-modelled information; and a move from a 'way of life' culture to a food business (BeVier and Lautner, 1994). A typical day on a medium-sized, 250-sow farrow-to-finish family farm at the end of the 20th century might have involved up to 15 hours of feeding, farrowing, breeding, vaccinating, castrating and tail-docking pigs amongst the dust and ear-splitting noise of a confinement operation (Donham, 1998). At the beginning of the 21st century, the average NZ dairy farm had 315 cows, seven times more than 50 years earlier. In this time, the stocking rate doubled and artificial fertilizer use increased, resulting in a 60% increase in production per cow and a seven-fold increase per farm worker. Over this period, there has been a change from nearly all farms providing all their own grazing and feeds to few now being so self-contained. There are 30,000 fewer dairy farms and this number is predicted to fall even further to achieve economies of scale (Ward, 2007). Similarly, the number of companies processing dairy products fell from 215 to 11 (Holmes, 1998).

These changes reflect increasing intensification and the drive for efficiency. However, they have not come without costs. The association of intensification with poor animal welfare, degraded physical environments and social inequity has become a major source of concern for some people. The intention of this chapter is to highlight the methods of modern intensification and some of the impacts on livestock, farmers and farm workers, and the environment, both physical and social. Finally, the relationship between economics and animal welfare is explored since it is this relationship that is at the forefront of many contemporary concerns with animal welfare.

Overcoming ecological constraints

The performance of any system, be it based on animals, plants or both, is dependent on, and constrained by, the availability of resources: soil, sunshine and water in the first instance. The efficiency with which solar energy is captured by plants and harvested and assimilated by animals, including man, determines the productivity of an ecosystem. The system is naturally constrained by factors such as climate and the physical environment. Typically, animal numbers are limited by the period of slowest forage growth, usually winter but also summer when summers are hot and dry and droughts are common. Changes in pastures, animals and management have long been used to overcome, or at least minimize, some of these factors; increasing inputs overcome ecological constraints (Table 3.1).

One of the most important factors affecting livestock productivity is the availability of feed; managing production depends, then, on controlling grazing by varying the frequency and intensity. Typically, and just as wild animals may disperse or migrate, or concentrate their activities to align with optimal forage availability, pastoralists and farmers control access to forage. For example, pastoralists expand herd numbers during rainy years so that a few might survive when drought returns. Similarly, in moving their sheep and camels north to pastures in the Sahara in the wet season, the Zaghawa of Niger follow separate parallel paths, leaving ungrazed strips for the return trek. And they restrict access to dry-season grazing, reserving forage for times when it is needed most (Durning and Brough, 1995). Increased growth and utilization of pastures and crops are a feature of farming. As livestock production intensifies, locally available forage tends to be replaced with traded cereals, oilseeds (e.g. soya) and processed by-products (e.g. fish meal), especially the protein-rich feeds and sophisticated additives that enhance feed conversion efficiency in the poultry, pig and dairy industries. The increasing use of grain for livestock feed is, in turn, enabled by the intensification of cropping resulting in increasing yields and declining prices, international prices halving over the last half-century (Steinfeld et al., 2006).

The means of intensifying animal production are legion. For example, Roche et al. (2017) summarized a century of changes in temperate grazing systems for dairy cows as including research into soil improvement, pasture breeding and grazing management to enable moving from providing cows with low-yielding, low-digestibility, unimproved pastures with

Table 3.1 A summary of some of the key inputs supplementing natural resources in modern farming (see Parliamentary Commissioner for the Environment, 2004).

Resource	Natural source	Supplementary source
Energy	Sun	Fossil fuels
Water	Rain, streams	Dams, wells, irrigation
Nitrogen	Fixed from air	Inorganic fertilizer
Minerals	Soil reserves	Mined, processed and imported
Pest and weed control	Biological, cultural, mechanical and some chemicals	Synthetic pesticides and herbicides
Seed	Some produced on farm	Production at separate locations
Management	Farmer and local community	Input suppliers, researchers and extensionists
Information	Gathered locally and regularly	Generic
Animals	Most produced on farm	Production at separate locations
Cropping system	Rotations and diversity	Monocropping
Varieties of plants	Thrive with lower fertility and moisture	Need high inputs to thrive
Labour	Family and hired	Hired and machines
Capital	Family and community	External sources

limited grazing infrastructure to highly specialized systems where highly selected cows graze productive and subdivided pastures. Intensification can be summarized as changes to the animals and to the farm:

Farm structural factors

- improving soil fertility by drainage and improving marginal land;
- fertilizers and irrigation to increase forage growth, e.g. synthetic nitrogen fertilizers and legumes to supply nitrogen;
- changing the composition of pastures, improved pasture management and better cultivars and making use of seasonal crops;
- conserving pasture (e.g. as hay, silage and balage) during periods of plenty for use during periods of relatively slow growth;
- increasing the utilization of pasture by managing grazing with rotations, strip grazing and technosystems;
- provision of shelter including housing; and
- increased farm sizes.

Animal management practices

- increasing stocking rates;
- artificial breeding and improved animal genetics and selection;
- breeding animals out of season or earlier in their lives, e.g. hogget mating;
- selecting desired genotypes such as those for superior meat yields, intramuscular fat and better growth rates;
- induced farrowing to synchronize births making it more efficient and economical to provide skilled supervision and enhance piglet survival;
- controlling internal and external parasites and predators;
- ultrasound pregnancy scanning to determine litter size, e.g. in sheep, so that nutrition can be better aligned with the ewes' requirements, a practice that has also reduced neonatal lamb mortality;
- shepherding sheep to ensure difficult lambing is assisted and orphan lambs are cared for;
- increasing feeding, e.g. dairy cows over winter when feed would naturally have been scarcer;
- more intensive monitoring to identify when health (e.g. parasites) and nutritional (e.g. trace elements) interventions or supplements are required;

Figure 3.1 Part of an advertisement for a veterinary pharmaceutical to treat and control internal and external parasites (*Photo courtesy Merial NZ, now Boehringer Ingelheim Animal Health New Zealand*). (Available in colour in the plate section between pages 178–179.)

- changing stock classes, e.g. heifer calving followed by slaughter after weaning, more bulls and fewer steers or heifers; and
- testing animals to determine their fitness.

Intensification, then, is an important and accepted part of farming, the development and marketing of technologies the focus of much agricultural research and many agribusiness interests (e.g. Figure 3.1).

While intuitively intensification might make sense, it is not without its problems. Increasing production by increasing inputs is not indefinitely sustainable, productionism criticized for its undeniable benefits being obtained at the expense of unintended consequences (Thompson, 1995; Zimdahl, 2006). A theme of Body's book (1984), *Farming in the Clouds*, is that the more that is spent on increasing production, the more economically inefficient it becomes; efficiency declines at higher levels of production. Firstly, if farms are already operating at a financial optimum, increased production may not cover the additional costs. The economics of marginal costs, or the law of diminishing returns, means that at some point the costs of extra productivity will not outweigh the extra time, money and resources required to generate them. It is a basic tenet of economic theory that it gets harder and harder to increase production as marginal costs and marginal production are inversely related. Secondly, the benefits of intensification must be balanced against other consequences, like changes to animal welfare. While some practices like determining litter size prior to birth have enhanced animal welfare (e.g. determining litter size by ultrasound scanning, Waterhouse, 1996), others may mean less space for animals, restricting their behaviour, and providing fewer facilities or resources. Fences, and especially barren and confined housing, limit animals' freedom to move and express normal behaviour, resulting in boredom and frustration. Less space means greater behavioural interactions, greater risk of parasites and diseases, and more effort being required to manage risks. Finally, intensification has also been associated with concerns as diverse as the loss of soil, biodiversity, water quality, deteriorating food quality, human health, family farms and communities and their values, changes to iconic landscapes, and rural poverty.

Farming systems: extensive and intensive

There are many farming systems throughout history and across the world, reflecting land use and intensification, economics a major driver of their development (Meyfroidt et al., 2013). Land use begins with that which is unexploited; some regions of the world are still characterized by their natural features, because they are either uneconomic to exploit or preserved for their aesthetic values. Increasing by levels of use, there is the land used by hunter-gatherers and nomadic pastoralists, then the land supporting extensive settled and intensive pastoralism, and cropping, and finally intensive, industrial or factory farming (Bray and Gonzalez-Macuer, 2010). Along this continuum, as natural ecosystems become farm land, there is a lessening of the retention of the natural features and an increase in the modification of the natural landscape, especially soils, water, flora and fauna. Environmental impacts include degradation and a loss of biodiversity, but an increase in the output of farming due to an increase in external inputs and management (Table 3.1). The diversity of farm animals, ecosystems and socio-economic systems or peoples means these systems are continuously adapting and changing, as they have since the beginnings of proto-farming. For instance, NZ's sheep industry is largely based on grazing improved pasture on land formerly native forest, grassland or wetlands. Since the 18th century, more intensive land use has enabled the replacement of native vegetation with introduced pasture species and improvement and optimization of different mixtures of species, the application of fertilizers to both overcome soil deficiencies and increase pasture production, and farm subdivision to more efficiently manage grazing. While the environment has certainly been modified, nutrition and animal health and welfare have been improved. Like any farming system, animals are managed and thus also have some of their behaviours restricted.

In today's world, extensive grazing or pastoralism, mixed farming and industrial or landless systems are easily and widely distinguished (Steinfeld et al., 2006). As they are so different, it is convenient to contrast extensive and intensive systems (Figure 3.2). Grazing and landless systems, the two ends of the continuum of land use, are both highly specialized. The former system is very highly dependent on the natural environment, requires minimal investment and yields are low. The latter system is less directly dependent on the natural environment (feed is purchased), requires high investment and is very high yielding. These industrial or landless systems are expanding and produce two-thirds of the world's poultry and half of the eggs and pork but little of the world's beef and other ruminant meats.

Figure 3.2 Examples of extensive and intensive or landless farming (*Photos: Mark Fisher*). Globally most protein is produced intensively with smaller amounts produced on grassland systems. All systems have inherent efficiencies and inefficiencies, yet their products are marketed and consumed similarly, making them susceptible to similar market forces. (Available in colour in the plate section between pages 178–179.)

Between the two extremes of land use are mixed production systems that have a more diversified range of products, depend on the natural environment and have a range of investment needs (Pradère, 2017). Most of the world's animal products are supplied through intensification, and will continue to be produced in intensive, large-scale commercial systems, because of the continuing benefits of economic concentration, favourable tariffs and subsidies, and not having to mitigate their environmental effects (Niamir-Fuller, 2016).

Extensive farming systems make up a quarter of the earth's ice-free land surface and are important to 200 million people. For example, pastoral systems involve moving animals, like cattle, yak, camels, horses, sheep and goats, to different areas as climate, environmental conditions and availability of forage demand, grazing being primarily dependent on range or natural grassland. Extensive systems are often in more difficult environments (e.g. arid and semi-arid zones, variable and harsh climates or steep, rugged or remote terrain) partly because they are unsuited to significant intensification. Typically, they have low inputs, management practices requiring much labour and specialized stockmanship, and lower productivity. Livestock usually get all or most of their food directly from their immediate environment and the animals, and the vegetation, are very dependent on seasonal patterns and vagaries in climate. Farming is often based on several species (e.g. goats, sheep and cattle) with diverse characteristics (e.g. strains and breeds adapted to different locales).

On extensive farms, the environment dictates the care of animals, including all or most of the nutrition they have access to. Some of the most significant features of animal welfare on more extensive farms relate to ecological, climatic, environmental and nutritional variation. Sometimes this means the animals suffer when climatic conditions are extreme, resources inadequate, or the animal unsuited to the environment (Fisher, 2011; Villalba, 2016). For example, recently shorn goats were susceptible to hypothermia when exposed to an unpredictable adverse weather event during autumn (McGregor and Butler, 2008). Those in poorer body condition, with less reserves or body fat, at shearing subsequently died, while those in better condition survived. Extensively farmed animals have an array of behavioural strategies to cope with and enhance their fitness, among them:

- spending more time foraging for high-quality food instead of eating abundant low-quality food, diet selection influencing where to feed, and reflecting competition, palatability and preferences, nutritional qualities, digestion and toxins;

- knowing the constraints to ranging, such as availability of water, predators and availability of shelter from weather;
- investing in fewer larger and healthy offspring (conversely, animals in better condition tend to have higher ovulation rates and subsequently give birth to, and successfully raise, more offspring);
- delaying breeding until more favourable conditions prevail;
- protecting young from predators;
- competing for resources;
- maintaining group sizes that optimize the use of resources and minimize the threat of predation and parasitism; and
- adopting a home range ensuring familiarization with the location of resources (food, water, shelter and escape from predators).

A rich behavioural repertoire is an essential part of adaptation, providing animals with the ability to adjust or adapt to their circumstances. The keeping of livestock in variable and varying environments utilizes the animals' flexibility of behaviour, enabling them to adjust and thrive. To do this, the animal must have both the means (the genetic potential) to thrive and a suitable environment (access to the resources needed to survive and be productive) and it must be managed appropriately. Infrequent or reduced human contact and intervention, and difficulties in husbanding animals in difficult and remote terrain and/or adverse weather, can have implications for the provision of husbandry and veterinary care, in relation to, for example, parasitic diseases and neonatal survival.

However, extensively farmed animals have relatively more freedom to decide about things that are important to them, the diverse environment providing opportunities for enhanced welfare. Furthermore, a greater degree of natural selection (and conversely, a relative absence of artificial selection and intensification) has seen animals better suited to their particular environments, with some remarkably resilient. For example, goats and sheep can go without water for several days in parts of Ethiopia and Kenya. Although an extreme situation, the Borana nomadic herdsmen may stop at a watering place for a period of time allowing their animals to hydrate before moving away to graze for 2–3 days without any prospect of water. The herdsmen aim to return to the source of water before the animals become dehydrated and their appetites suppressed. Finally, and often because they are in remote locations, extensively farmed animals are exposed to predators (Laundré, 2016). Feral pigs are one such threat to young lambs in NZ (Figure 3.3) and, as well as coyotes, cougars prey on livestock in the western US. In Norway, sheep grazing freely on open forested and alpine summer ranges are subject to predation by wolves, brown bears, wolverines, golden eagles and lynxes.

Figure 3.3 Feral pigs scavenging on the margins of farmland in NZ (*Photo: Mark Fisher*).

An overriding feature of extensive farming is that, in general, it appears to be in keeping with the nature of the animal, it leading a relatively natural life, growing and reproducing normally. Except for painful husbandry procedures such as castration and dehorning, and the emotional stressors associated with weaning and handling, it is suggested that the animals suffer little and pretty much experience the normal pleasures of life. However, while many believe that the welfare of animals is acceptable in extensively farmed systems, they also hold that improvement is required (e.g. Doughty et al., 2017).

There are two general approaches to enhancing animal welfare. The first is to enhance the type of animal so that it is best suited to cope with its environment: fitting the animal to the farm. For example, selection for traits important in maternal and grazing behaviour, and for disease resistance. While animals must have suitable traits to survive in a particular habitat, they must also be able to adjust their behaviour to suit the conditions. The second approach involves ensuring that the physical environment provides the resources for the animal to cope: fitting the farm to the animal. This includes providing shelter, improving handling procedures, controlling parasites, giving supplements and missing trace elements, and ensuring stockmen have the necessary skills and expertise.

There are many different management practices that are used to enhance the fitness, welfare and productivity of farm animals. They range from breeding animals with superior traits to ensuring that only younger, fitter animals are retained in challenging environments, to keeping different generations of animals in the same environment, harvesting only a proportion of any cohort to ensure subsequent generations of, for example, the flock collectively retain knowledge of the environment. A combination of artificial and natural selection means that some breeds and strains are better adapted to extensive production systems than others. For example, the Herdwick, Swaledale, Welsh Mountain and Scottish Blackface sheep have developed to survive the harsh conditions commonly encountered during winter in the UK, while Aubrac, Gascon and Saler breeds of beef cattle are known for their hardiness in France. Among the many breeds of goats in Africa, the Swazi is well adapted to subtropical and sub-humid areas, the Sudan Desert breed to semi-arid and arid regions, and the West African Dwarf goat to humid forest zones.

Intensive, landless or industrial systems have transformed animal production, so much that it can be asked if they are still regarded as farming. The foundation of good stockmanship and the ancient contract is displaced by technology and the drive for efficiency. The more intensive systems, pigs and poultry, aim for rapid production using standardized methods and a high degree of mechanization and automation. Farm animal numbers tend to be large, animals and systems highly productive with relatively few diverse characteristics and founded on an economy of scale. For example, modern broiler or meat chicken production systems consist of multiple million-dollar sheds each housing tens of thousands of birds. The six- to seven-week growing cycle is managed by few staff with most of the work, as well as lighting and heating, automated. Significantly, the most intensive modern farming systems are most likely to be associated with concern for less than optimal animal welfare.

Welfare problems associated with some of the more intensive systems have been well documented and some will be considered in greater detail later in this chapter. However, in the first instance, many relate to close confinement, unforgiving surfaces and barren environments with lack of opportunities to satisfy physical and social needs: for example, hens who are motivated to roost during darkness being deprived of perches (Olsson and Keeling, 2002). High stocking rates can result in fighting, injuries and deaths. Like animals in less intensive systems, they are subject to painful husbandry procedures like tail docking and castration, transport and slaughter. Finally, highly productive animals can be susceptible to production-related diseases. For example, sows with large litters inevitably have to produce more milk, and, when unable to support the litter, the farm requires stockmen with greater cross-fostering skills (Levis, 2016).

Stockmanship

The aims of good animal husbandry are to ensure animals have the resources they need to be comfortable, fit and feeling good, intervention generally only being required when something is lacking. Stockmanship is fundamental to animal welfare and the attributes of good stockmanship include:

- Experience and learning: individuals typically draw on a lifetime of practical personal experiences with animals and farming such that stockmanship becomes second nature, where 'feel' and experience are valued as much as specialized knowledge and measures.
- Personal qualities: patience, empathy, and so forth, are traits or attitudes considered necessary when working with and being responsible for animals. While the personal qualities of a good stockman have been described in the scientific literature, the value of experience has not been so well addressed. In part, this reflects the scientific study of stockmanship being based on modern pig, poultry and dairy cattle farming, systems that are quite intensive and removed from the ecological foundation characteristic of more extensive farming systems.
- An understanding of the constraints and opportunities afforded by the physical environment – the climate, the terrain and the biota – experience with animals going hand in hand with knowledge of the local land and climate.
- Knowing the normal behaviour of animals and, being observant, having the ability to recognize and deal with abnormal behaviours. For example, the first sign of housed sheep experiencing ill-health was that the animals did not look the stockman in the eye as he walked past the pen (Smart, 2004).

The knowledge and attitudes of good stockmen and women, developed and valued by individuals as a vocation, are generally unavailable to professionals and others outside the industry, although they play important roles in contributing to parts of that knowledge. This holistic nature of stockmanship ranges from knowledge of the link between a person's nature and an animal's response (e.g. a poor stockman can turn a good cow bad), to the link between an animal's performance and the constraints of the physical and climatic environment. Furthermore, it is individuals, be they shepherds, herders or others, who are best placed to increase their tacit and local knowledge in ways which authorities, teachers and others cannot. It is these people who must ultimately decide which, if any, advice or guidance to follow in their circumstances, a skill in itself.

Stockmanship is based on a respect for the essence of the animal based on stewardship (Gatward, 2001). Care for animals is a cornerstone of animal husbandry, evident in traditional animal husbandry systems. In contrast, very intensive systems resemble recipe-like farming, an example of 'McDonaldization': the principles of efficiency, calculability, predictability, conformity and control of fast-food restaurant management that are coming to dominate other sections of society (Ritzer, 2004). While the attitudes of people working with animals play an important part in enhancing animal welfare, stockmanship, especially in more extensive and pastoral systems, is a relationship, not simply the behaviour of people working with animals (Boivin et al., 2003; Hemsworth, 2004; Hemsworth and Gonyou, 1997).

Good stockmanship means:

(1) Always ensuring animals have adequate reserves, or that there is adequate resilience in the farm system, to enable animals to cope with their changing and variable environments.
(2) Ensuring animals are genetically adequately suited and acclimatized to the environment and the production system.
(3) Valuing and using the practical knowledge and experience people have of the interaction between the environment, the animals and people.
(4) Keeping the environment, the climate, the animals and farm management expectations in equilibrium.
(5) Providing the right environments, resources and management for animals in order that they can adapt.
(6) Attending to or minimizing those aspects known to have a risk for animal welfare, and that are controllable, so that animals can direct their resources or adapt to the stressors beyond the control of farm management.
(7) Facilitating and encouraging those with responsibilities for the care of animals to have the time, opportunities and confidence to achieve these aims.

Some of these may be less important in more intensive systems where some ecological constraints have been removed and the environment constrains or prevents others from being addressed. However, more highly performing farm systems require greater skills of stockmanship and animal husbandry.

Performance

Farm animals are useful because of their productivity: the more animals produce, generally the more humans benefit. High farming, the specialized, progressive and increasingly capital-intensive, high-input high-output, industrial-like systems that are now a feature of much of the modern world's food production, has seen spectacular increases in food production. The average dairy cow in Denmark produced 1000 kg of milk in 1860, a mere eighth of what it produced in 1999 (Bielman, 2005; Sandøe et al., 1999). The progressive technological achievements of modern farming accompanying those higher yields are equally impressive (Figure 3.4).

Figure 3.4 Driven by the need to boost labour productivity, the milking machine has been one of the most innovative technologies in modern dairy farming (Bielman, 2005). In less than a century, dairy farming has gone from milking by hand, e.g. in the 1930s–1940s (*Photo: Alamy*), to machine and robotic milking (*Photo: Mark Fisher*), requiring an increase in herd size to cover the investment. The yearly labour needed to 'manage' a cow has dramatically reduced, from 330 man hours in the 1950s to about 40 in the 1990s.

The wild European boar, from which the modern domestic pig has been partly derived, usually has one litter per year of two to five young whereas modern sows can have ten or more, 2–2.5 times a year (Arey and Brooke, 2006; Bezant, 1999). Modern layer hens produce over 300 eggs per annum, whereas the jungle fowl, believed to be an ancestor of the domestic chicken, lays some three or five to seven eggs per clutch, maybe a few less in any re-nests (Collias and Collias, 1967) and domestic hens allowed to become feral perhaps two clutches of 14–15 eggs each (Duncan et al., 1978). Chicken meat production has made perhaps the most progress, for example, the modern broiler is three times heavier than the junglefowl at hatching (Siegel et al., 2009). Between 1957 and 2001, broilers or meat chickens became up to six times heavier, yielding up to 10% more breast meat, and were grown or finished in less than half the time they once were (Havenstein et al., 2003). Finally, the modern dairy cow produces some 2–12 times more milk per lactation than the beef cow in raising a calf, an animal more similar to the wild or natural ruminant from which our modern cattle breeds have evolved (Holmes et al., 1987). These examples of increased production have been most sustained in the last half-century (Figure 3.5).

The enhanced yields are a culmination of breeding and selection, better nutrition, protection from predators, protection from and treatment for the effects of parasitism and ill-health, mechanization and improved managerial skills. As well as selecting for more productive animals, more efficient individuals, those able to produce more meat, milk and eggs with less feed, are increasingly being identified. Feed conversion efficiency reflects the energy required to rearrange the components of a diet (fatty acids, amino acids, lipids, minerals, vitamins etc.) into fat, protein and lactose in milk, or fat, protein and bone in growing animals. Altering the animal's nutritional regulatory mechanisms is not new; selection for high production may have altered the mechanisms regulating appetite, overriding the normal regulatory mechanisms preventing rapid growth or excessive fatness (Rauw et al., 1998). And some of the impacts may seem unrelated, for example, rapid growth in farmed fish affecting the structure of the inner ear, important in hearing and maintaining balance (Reimer et al., 2016, 2017).

Hard work and undesirable traits

While many have catalogued the harms animals can experience or be exposed to, what harms are associated with their high productivity? How hard do farm animals work? By using the convention of energy exchanges per unit of metabolic body size, it is possible to compare energy taken in

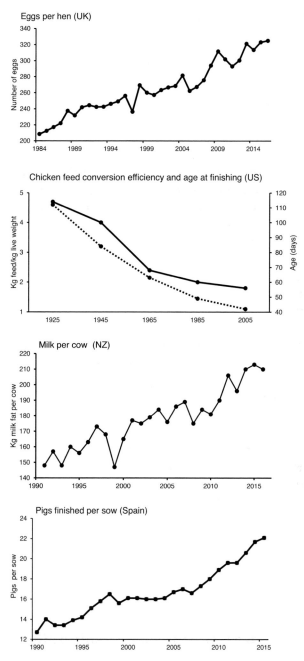

Figure 3.5 Some of the recent advances in production from farm animals. The average number of eggs laid yearly by hens (top panel), feed conversion efficiency (solid line) and age at slaughter (broken line) in chickens (second panel), annual milk production (third panel) and pigs finished per sow (lower panel). Note the different scales (years) on the horizontal axes.

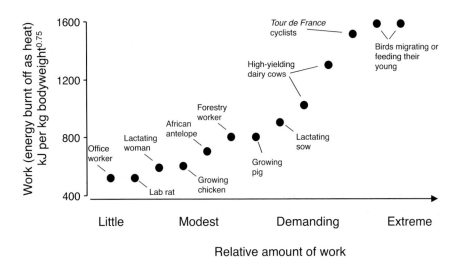

Figure 3.6 The relative amount of work humans, farm animals and wild animals perform (Webster, 1992). A relatively sedentary adult laboratory rat or an office-working man, for example, each consume, allowing for their body weight, 520 kilojoules of energy per 24 hours. This is used for maintaining body functions and any exercise and burnt off as heat, altogether a relatively little amount of work.

as food, and burnt off as heat, in animals of different sizes (Figure 3.6). Harder-working or higher-producing animals, and humans, require more food and produce more heat.

Undertaking greater physical activity (e.g. foraging ruminants such as antelopes) means more work, the most extreme examples being birds migrating or feeding their young. Whilst the work involved in growth amongst farm animals is relatively modest, lactation in high-yielding dairy cows can be very demanding, the equivalent of a human jogging six to eight hours a day, every day, for most of the year. The dairy cow's sustained energy demands are exceeded only by the frenetic and transient activities of birds migrating or feeding their young, and by Tour de France cyclists.

Not surprisingly, some farm animals are predisposed to ill-health requiring increased stockmanship and management skills if they are not to suffer or require culling. Undesirable behavioural, physiological and immunological traits and production diseases are usually linked with high levels of productivity and/or high-performance systems (Millman et al., 2000; Ott, 1996; Rauw et al., 1998). Where once natural selection may have removed pathological traits from a population of animals, artificial selection and management means they are tolerated, sometimes favoured. For example, some such as spectacular growth rates may be indicative of damaged hypothalamic mechanisms regulating satiety, so that hunger is not reduced, leading to overeating (Burkhardt et al., 1983).

Production diseases are man-made, the combined strain of high production and modern husbandry resulting in metabolic breakdown, often a necessary outcome of modern farming systems (Mulligan and Doherty, 2008). Enhanced production, often combined with inadequate housing and genetic or nutritional management, brings the risk of failure of the animal's physiological system to cope with the demands of high production and modern intensive husbandry (Chriel and Dietz, 2003; Mulligan and Doherty, 2008; Nielsen, 1999; Nir, 2003; Payne, 1972). The selection and management of animals for 'efficient' agricultural production can result in the breakdown of the various metabolic systems, often requiring greater husbandry intervention including the routine use of medication. The most well-known production diseases are metabolic; the degree of metabolic load that distorts normal physiologic function is, in turn, a function of the intensity (extent to which supply fails to meet demand) and duration of the load. The nutrient supply for normal metabolism comes from diet and body reserves but with increasing load, the animal's system is challenged resulting in behavioural and physiological adaptations. If the load is too much then the system malfunctions resulting in disease, or, if severe or prolonged, in death. Production diseases are thus caused or amplified by systems of management, especially feeding, and breeding high-performing animals in which production exceeds the ability of animals to cope. While high production and good welfare are not necessarily antagonistic as long as quantity and quality of input are able to match high output, selection for greater yield will increase the risk of the animal being unable to cope or adapt.

Production diseases can be grouped as caused by or associated with genetic, nutritional or farm management practices.

Genetic predispositions include:

- body conformation or size (e.g. double muscling in cattle);
- growth rate, feed conversion efficiency, and high milk or egg production (e.g. skeletal disorders in broilers); or the result of
- inadequate culling of undesirable genes (e.g. lameness in pigs).

Nutritional conditions may comprise:

- imbalances in diet composition because of over-attention to feed digestibility (e.g. pasture monoculture and bloat in dairy cattle);
- demands for production exceeding physiological reserves (e.g. egg production and osteoporosis); or
- providing insufficient feed (e.g. hunger in pigs and breeder poultry).

Husbandry practices can consist of:

- high stocking rates and reduced locomotion in indoor systems (e.g. lameness in broiler chickens);
- group size (e.g. social instability in large dairy herds);
- adverse climates (e.g. death from hypothermia in lambs);
- prolificacy compromising survival (e.g. increased deaths of triplet lambs);
- inadequate facilities, equipment, procedures or staffing (e.g. milking equipment or technique and mastitis in dairy cows);
- inadequate exercise, or insufficient space (e.g. osteoporosis in caged layer hens).

While not just associated with the most intensive systems, production diseases are most apparent and have been well documented in the most intensive farming systems involving poultry, dairy cows and pigs.

Selection for increased production has not surprisingly increased birds' appetites. Broilers or meat chickens become obese if not slaughtered around seven weeks of age for example. However, the parent stock, those mature birds who mate and lay to produce the chicks for the broiler industry, have similar appetites and their intake must be significantly restricted if they are to maintain health and reproductive competence. Restrictions, either by feeding every second day or by limiting the feed each day, can be 60–80% less than what the birds might want during rearing, and 25–50% less during laying. These feed-restricted birds, half the size of those fed to appetite, consume their ration in minutes and consequently show behaviours indicative of boredom and frustration (e.g. aggression, pacing, stereotypical pecking at empty feeders and over-drinking) and appear chronically hungry. Compared with birds fed to appetite, feed-restricted birds worked four times harder to obtain food after being deprived of it for 72 hours (Kjaer and Mench, 2003). Faster-growing birds tend to have bones that are inadequate to support their greatly increased muscle mass, resulting in skeletal disorders. Foot, leg and locomotion issues can affect 0–90% of broilers, many having a detectable gait and some suffer from abnormalities suggestive of compromised welfare. Birds at the end of their laying cycle can have higher levels of bone fractures and other injuries or deformities (Kestin et al., 1992; ; Knowles et al., 2008; Nasr et al., 2012; Sandilands, 2011; Shim et al., 2012; Thorp, 1994; Wilkins et al., 2011).

The incidence of some of the more common metabolic production diseases in dairy cows is generally low, less than 10% of cows being affected per year or lactation (Pryce et al., 2016). They include undernutrition (reflected in poor body condition or excessive loss of body condition), milk fever (a calcium imbalance) and ketosis, an excessive demand for glucose (Lescourret et al., 1995; Pösö and Mäntysaari, 1996; Pryce et al., 1997). The high-producing dairy cow may also suffer from hunger, discomfort from poor housing, and metabolic or physical exhaustion after prolonged high production (Webster, 1994). Mastitis, lameness and infertility (e.g. unobserved oestrus) are commonly associated with modern dairy farming. There is, quite literally, a mass of scientific papers describing the complex antagonistic relation between high-producing dairy cows and their health, welfare and fertility, with much still to be clarified. For example, a fifth of cows within some UK herds had mastitis or were lame (Amory et al., 2008) and there was a 1% decline in calving rate to service every three years over the last three decades of the 20th century (Darwash and Lamming, 1997). Increased milk yield is associated with the cow using her body reserves, even when sufficient food is supplied. Lower body condition contributes to infertility as most animals only usually breed when they are in adequate condition. Veterinary pharmaceutical intervention may compromise welfare, preventing the cow from using one of her few coping mechanisms, that is, delaying the next pregnancy. Reproductive failure is a major reason for culling and much has been written about the worldwide poor, and declining, reproductive performance of the high-producing dairy cow (Lucy, 2001; McDougall, 2006; Dobson et al., 2007, 2008; Walsh et al., 2011). Failure to be able to meet the high-producing cow's demands for feed and to provide an adequate environment have left her susceptible to disease, inadequate nutrition, poor housing and so forth, with animal husbandry unable to keep pace with her modern genetics (Dobson et al., 2008). The association between production and disease may also be evident in an association with culling rates (Agger and Willeberg, 1991; Esslemont and Kossaibati, 1996; Compton et al., 2017). While there are many causes of death and culling, the most common disorders (udder and limb deformities, poor reproduction, and calving, intestinal and metabolic problems) are all multifactorial, related to feeding, the environment, management and productivity. For example, the proportion of mature cattle (assumed to be mainly dairy cows since they were the largest proportion of the national herd) that were rendered has steadily increased in Denmark. In 1960, there were just over 200 'deaths' per 10,000 cow years, the equivalent of 2% of the dairy cow herd. By 1980, it was 4.3%, the increase suggesting an association between production, disease and mortality (Agger and Willeberg, 1991).

Finally, there are examples of pigs displaying behaviours indicative of not being able to cope or adapt (Van Putten, 1988). These include:

- vulva-biting, endless fights and death because of an inability to interact properly in groups;

- being unable to wallow to keep cool;
- stereotypic or pointless, repetitive behaviour reflecting an inability to express normal behaviour in confined and barren conditions;
- concentrate feeds, while satisfying hunger, leave pigs foraging, their urges unsatisfied, resulting in sham-chewing, rooting and exploring after feeding;
- the reduced space of farrowing facilities preventing the sow from removing piglets from her lying site;
- lack of resources and opportunities for nest-building can result in prolonging farrowing, and sows being afraid of, and attacking, their piglets; and
- the lack of exposure to an enriched environment and routines means pigs do not develop the ability to deal with varied and/or stressful situations, such as transport.

Another consequence of intensification is the risk of piglet deaths, litter size linked to piglet mortality. The intense competition for teats and the likelihood of hunger mean those of low birth weights are most susceptible, piglets dying from starvation, chilling and injuries (see Fraser, 2014). Like most of the problems highlighted in this chapter, efforts are being made to mitigate them including piglet health and vitality selection programmes, sow nutrition and minimizing stress during pregnancy. Management interventions to deal with increasing litter size include split suckling, cross-fostering, nurse-sow systems and various weaning and rearing strategies (Baxter et al., 2013; Rutherford et al., 2013).

As well as production diseases, it is worth drawing attention to zoonotic diseases (Leibler et al., 2009), infectious diseases in human populations having their origins in animals, mostly farm animals. The most intensive production systems, where large numbers of animals are raised in confinement with high throughput and rapid turnover, are novel ecosystems. These environments may be predisposed to the evolution of novel pathogens and their transmission to humans, providing biosecurity and health challenges. And finally, there is the link between animal welfare and food safety, something intuitively accepted but rarely investigated. It is well known that stress, including depriving animals of feed, transportation or heat stress, can interfere with gastrointestinal function, increasing the risk of the establishment of foodborne pathogens in animals. For example, improving health and welfare could reduce risks of reduced food safety in pigs and chickens (Alpigiani et al., 2016; Bull et al., 2008). However, and probably reflecting the complexity of farm systems and the multitude of factors affecting gastrointestinal microbiota, there is little conclusive evidence of differences between different animal production systems (Lara and Rostagno, 2018).

How improved are improved farm animals?

Increased productivity appears to have come at the cost of changes in traits, management and the environment: loss of fitness particularly evident in the increased behavioural problems, ill-health, infertility and production diseases seen in modern farm animals such as the broiler or meat chicken and the dairy cow. This is not an exhaustive or comprehensive review (see Rauw et al., 1998), just a selection of some of the undesired characteristics of high-yielding animals. Not all animals experience these traits; some individuals and populations will be better, others perhaps worse. And the symptoms or indicators of welfare described are also a window, they change over time. The point is that it is the systems that have helped produce or have resulted in these conditions, at some time in our history. Similarly, these are not a feature of all farming systems though many or all systems are under pressure to become more intensive. The point is that there are, or have been, negative effects of farming animals to yield high amounts of meat, milk and eggs. Some are being addressed in breeding and selection programmes and animal husbandry, for example, in dairy farming (see Chagas et al., 2007). Similarly, the poultry industry is addressing the need for slower-growing birds, more space, natural day/night lighting patterns, environmental enrichment and addressing public health concerns (Swormink, 2017). And not all selection programmes aim to increase productivity; some, such as resistance to disease and parasites and tolerance to cold, also contribute to enhanced animal health and welfare (Lawrence et al., 2004; Scobie et al., 1999, 2007).

However, production diseases may also represent a trend towards the breeding and farming of animals that are progressively unfit for purpose. John Webster, a veterinarian and one of the notable figures of modern farm animal welfare, has written in *Animal Welfare: Limping towards Eden* (2005a):

> *When modern genotypes of farm animal bred for intensive systems are examined ... the breeders' claims to genetic superiority come apart at the seams. Broiler chickens that are crippled in large numbers or die from cardiac failure if reared beyond the conventional slaughter date of 42 days are clearly not fit. The food intake of adult broiler breeders has to be severely restricted to stop them eating themselves to death. Breeding sows so large, clumsy and/or lame that they have to be confined in farrowing stalls to prevent them from crushing their piglets as they crash uncontrollably to the ground are not fit. Dairy cows that have to be culled for reasons of infertility, mastitis or lameness after less than three lactations are not fit. In these examples the artificial selection of food animals for traits dominated by the quality and efficiency of food production per se has imposed a severe cost largely expressed as a loss of fitness in the breeding generation.*

Intensification can have undesirable side effects and require enhanced husbandry skills, resources or effort to realize the potential of the farming systems and to address any adverse impacts on animal welfare. If it is becoming progressively harder to meet the needs of high-performing animals, we might also ask, what impacts is it having on farmers and their farms?

Impacts on farmers

Hard work and long hours are, and have always been, part of farming: 'Who first invented Work – and tied down the free' (Charles Lamb, 1775–1834). For example, 16- and 17-hour days are not unusual around calving time on modern dairy farms. However, the extent seems to have increased, evident in this comment from an elderly farmer: 'In the old days a dairy farmer could milk 280 cows, have a good living and still have time to talk to the neighbours. Today we work harder and harder for less money' (Anonymous, 2012a). In NZ, nearly two-thirds of the dairy farm population work more than 50 hours a week compared with only a quarter of NZ's general workforce (Clark et al., 2007). And while only 10% of the NZ general workforce work 60 or more hours a week, a staggering 40% of dairy farm employees, 45% of their employers, and 49% of those self-employed and without employees, worked over 60 hours. Even 18% of unpaid family members worked these long hours (Wilson and Tipples, 2008). And it is not just the long hours, but also the need to continually improve efficiency and productivity, to intensify, which is affecting farmers. Take for instance this reflection from a farm advisor:

> *The farmers gave us a clear message that they've had a gut's full of continued prodding to produce more and more beef per hectare. They reminded us about two really important points. The first is that many have already reached the environmental limits of their soils – more hooves per hectare would be counterproductive, they say. The second is that they did not have the physical capacity or motivation to work any harder – they were already stretched in their capacity to get the jobs done and they were not getting any younger. (Thomson, 2009)*

Farming is also increasingly financially difficult, at least for some. In the US, off-farm income has become important, especially for young and beginning farm households, not only in supplementing farm income, but also in reducing the household to exposure to the volatility of farm income and giving better access to credit. The traditional way for young people to progress into dairy farming in NZ was to begin as a farm worker and progress to become a herd manager, farm manager or share-milker. A share-milker owns and milks the cows and shares the profits with the landowner and as many as a third of dairy operations involve share-milking. This financial system allows young people to work their way up by building up assets in the form of cows, some of which could be sold to provide the deposit on the purchase of a farm. In 1978, when the average dairy farm was 64 hectares and the average herd 118 cows, a share-milker might have needed to sell 200 cows. A mere 30 years later, land prices and farm and herd sizes had increased so much that it required some 1000 cows to provide the deposit, excluding many from farm ownership and favouring equity partnerships and joint ventures, or people developing their careers as farm managers without seeking farm ownership (Anonymous, 2012b). The volatility of farming incomes, evident for example in global dairy prices, is a further challenge many farmers face.

Farmers, farm workers and their families are exposed to a range of social and environmental stressors, particularly the climate, and the expectations and pressures 'imposed' by society. For instance, an animal's welfare, perhaps once solely the domain of those charged with care of the animal, is now determined by society, increasingly laid out in legislation, assurance schemes and guidelines, to which farmers and others are expected to conform. Increasingly, farmers' knowledge and skills, practices, and their animals and facilities are being monitored, audited, verified or inspected, activities that impinge on farmers in different ways. While perceived as necessary and inevitable, inspections were also considered unfair and inconsistent and created uncertainty (Anneberg et al., 2012). Even the responsibilities of determining animals' breeding objectives – of ensuring animals are fit for purpose on behalf of the wider industry – can be daunting.

As mixed or diverse family farms have shifted to larger, specialized operations, there are fewer people farming. Historically, farm labour was based on families, and a proportion of young and fit workers, wanting to progress into management and farm ownership (Trafford and Tipples, 2012). However, attracting and retaining skilled farm workers is now a problem in many countries for the following reasons:

- changing ownership structures, e.g. corporate farming, mean less access to family labour;
- declining rural communities offer fewer educational, social and support opportunities for potential farm labourers; and
- an image of poor employment conditions borne of long hours, demanding rosters, and treatment of staff making them susceptible to stress and fatigue, illness and injury.

Technology rationalizing and mechanizing work on the more intensive farms has resulted in many jobs becoming unrewarding and dehumanizing, especially when menial, repetitive and tedious, and it does not matter if they are undertaken in a mediocre fashion. The satisfying skills of animal husbandry and stockmanship are being replaced by production processes planned and maintained by managers working according to 'recipes' and computer programmes. I am reminded of one chief executive who demanded to know why livestock were not being grazed on a block as he had planned, only to be informed it was covered in snow. Increasingly, such approaches mean communities will lose the skills and experience of good stockmanship. Farming requires individuals having the freedom to operate and develop sustainable environments, risk management systems, and the implementation of industry-agreed standards, including animal welfare, within their own ecological and social contexts. Alternatively, if systems like dairy production are to become so complex that they require multidisciplinary inputs, for example, the farmer, veterinarian, other farm and financial advisors (Mulligan and Doherty, 2008), should the farmer bear all society's risks?

While changes in farming have been well documented, less obvious have been the changes in social and community life. However, two seminal works, *As You Sow* (Goldschmidt, 1947) and *Pigs, Profits, and Rural Communities* (Thu and Durrenberger, 1998), and others (e.g. Lobao and Stafferahn, 2005; Pew Commission, 2008), describe the social and environmental impacts of agricultural industrialization in the US. They suggest that industrial farming emphasizes technology and agribusiness over people, with the costs often borne by many and the profits shared by few. The contributors to *Pigs, Profits, and Rural Communities* note several concerns. Firstly, the health demands of exposure to fine airborne dust containing bacteria and bacterial products from dried animal faeces and feedstuff, and ammonia and hydrogen sulphide from animal urine and faeces. These bioaerosols and hazardous gases can be a health hazard for both confined animals and humans working with them. Intensive pig farm workers experience more major chronic respiratory symptoms, occupational asthma and obstructive lung disease than other people. The adverse health effects are also evident in the mood disturbances of those living nearby. The odour is more likely to make people angry, depressed, tense, fatigued, confused and less vigorous, and prevent people from being positive. Secondly, environmental pollution from ammonia and spills and seepage from waste lagoons present a significant hazard. Excessive nitrogen can change or damage habitats, contaminate groundwater and lead to eutrophication and to mineral build-up in soils. Thirdly, traditional local farm labour, associated with smaller, more diversified operations, tends to be replaced with underpaid, transient labourers. For instance, around 20 local jobs are lost for every large-scale unit with a 10,000-animal facility employing three part-time employees on a minimum wage. In turn, this leads to community changes such as instability in schools, stress in education and health care, demand on food banks, and increases in crime and the need for more enforcement and judicial systems. Finally, community relationships can be disrupted. Public policy favours interest groups having to establish a high standard of proof in objecting to industrial farming, rather than farm industries having to demonstrate their safety. In addition, food production is becoming further concentrated in the hands of fewer people, with democracy undermined by political biases and corporate or big business influences, and communities becoming less viable and more inactive because there are fewer people spending money locally.

All these changes were described over 70 years ago in *As You Sow*, a study of the effects of agribusiness in California. This work compared social relationships and the quality of life in rural communities surrounded by either small (average 57 acres) or large (average 497 acres) farms in 1944. Though having fundamentally similar cultural heritages and economic circumstances, the social conditions in the towns of Arvin and Dinuba diverged due to the effect of scale of the farm operations upon the character of the rural community. Surrounded by large-scale farm operations, Arvin had the poorer social environment. The study inferred a causative relationship between agricultural industrialization or corporatization (reflected in farm size) and the urbanization of rural life. As average farm size increased, the number of people supported in the rural area and local community declined, profitability and community virtues being mutually antagonistic. The effects highlighted in Goldschmidt's classic study are evident in more recent works considering efforts to save smaller, independent farming units within the US (e.g. Comstock, 1987; Strange, 1988). These and other studies recognize the social costs of increasing farm production: food production and power concentrated in the hands of fewer people undermining the health of the democratic political system. The loss of people working diverse farms also means a loss of knowledge, often embedded in communities and their histories. Such knowledge borne of regular interactions between animals and farm ecosystems is important in adapting to changing environments, in guiding the relationship between humans, animals and plants and the land, and in caring for animals.

Given that intensive farming is difficult for farm animals, and for farmers and their communities, why persevere with it? What drives people to select animals that consume less food but grow faster, to feed and care for high-performing animals? The answer is simple: intensive farming is very successful and brings a range of important benefits. Firstly, it provides food and fibre; farming is more effective than hunting. And, increasingly, fewer farmers produce food for the rest of the population, enabling the human population to increase, while also releasing them from the demands of growing and harvesting. Secondly, it is a commercially successful strategy. The need to be profitable is one of the fundamental goals of agriculture (along with safety, the environment and in keeping with a just social order). But, it is well to remember that a proportion of farmers, perhaps 20–50% depending on the industry, have no interest in intensifying. For instance, only half of deer farmers surveyed had, in the previous three years, made a change to a farming practice or adopted a new technology, mainly to increase productivity, profitability or improve animal welfare (Deer Industry New Zealand, unpublished data).

In general, farmers tend to use their resources (land, soils, pasture, water, labour, technology) and management to make a profit subject to the constraints imposed by their, and society's, preferences and constraints. An overriding feature of the farming system is that productivity has increased, not only when viewed over the last 10,000 or more years, but also especially over the last half-century. There are many interrelated and interdependent motivations for, or drivers of, intensification, and many enabling or facilitating factors (Drabenstott, 1995; Firbank et al., 2013; Fraser, 2005; Steinfield et al., 2006; Van Vliet et al., 2015). They include economic, technological, institutional and farm characteristics, as well as changes in demographics, and sociocultural globalization and urbanization. For instance, the main drivers of land use intensification between 1945 and 2013 were technology (45%, e.g. mechanization), institutional (45%, e.g. production subsidies), farm and farmer characteristics (40%, e.g. motivation for farming), location factors (31%, e.g. soil quality) and economic (28%, e.g. globalization of agricultural markets). Drivers cause changes in the farming system, either directly (proximate) or indirectly (underlying or ultimate). At the root of intensification are consumer demands and an array of technologies and management tools enabling producers to meet those demands (Drabenstott, 1995).

The direct drivers of intensification are farmers' goals (Parminter and Perkins, 1997), predominantly financial or lifestyle. They include building a valuable business, producing to maximize farm profits, looking after the welfare of livestock, creating increased opportunity for other farmers, paying off debts, maintaining a stable farming system, having time for family and friends, being self-reliant in decision making and having variety in their work. The indirect drivers are demographic, economic, sociopolitical, scientific and technological, and cultural and religious; they are fundamental societal processes that produce the direct drivers. Although there are many factors, a number of themes are apparent. First is the demand for products reflecting human population growth, rising incomes, and urbanization. Demand for meat, milk and eggs has increased as consumers have become more affluent; global economic activity has vastly increased as incomes have doubled between 1950 and 2000. Livestock provide over half of the protein supplied in industrialized countries, but less than a quarter in developing Asia, the Near East and sub-Saharan Africa for example. Conversely, economic stagnation and decline brings a decline in consumption. As income rises consumption also changes; fat, meat, fish, fruit and vegetables become more important with staples such as rice, wheat and potatoes less important.

Secondly, the motivations for intensification may have included the attraction of using modern 'scientific' methods, the benefits to animal health welfare of practices such as removing animals from contact with their faeces, and the use of prophylactic treatments such as vaccines and antibiotics. These would have been reinforced by contributing to economic development and social progress, farmers taking pride in large yields and greater fecundity. This paradigm has dominated farming over the last half-century, it making good sense to grow and produce more food. Facilitated by modern science and technology, and adopted by willing farmers, the impacts of improved genetics and increased fertilizer and pesticide use were slow to be questioned.

Thirdly, the desire or need of farm businesses to remain or become more profitable, to extract financial benefits from farming, is one of the main factors encouraging or demanding intensification (Small et al., 2005). Farming needs to remain profitable in the face of, for example, increasing land values, declining market prices, marginal returns on investments and rising national currency exchange rates. Improving production is motivated by the need to absorb or reduce costs, increase returns and, in some countries, access public payments. While farmers may increase production, they acknowledge environmental and animal welfare concerns and the impact of intensification on the perception of their industry, both locally and internationally (Food Ethics Council and Pickett, 2014). For example, a common means of intensifying sheep farming is to increase the number of lambs born, but this can conflict with the goal of maintaining animal welfare. Like any business, margins between incomes and expenses are continually squeezed, requiring operations to get bigger to take advantage of economies of scale, grow animals faster, produce niche or high-value products, and spend more time learning how to do things better: to 'get smarter or go broke'. Other options include accepting a lower income, reducing costs, increasing off-farm income or selling up and investing elsewhere. Productivity is a key driver of intensification: 'to make two ears of corn, or two blades of grass, to grow upon a spot of ground where only one grew before' (Swift, 1726). Generally it makes good economic sense if the returns from the additional produce cover the costs involved in producing it.

Fourthly, intensification is enabled or facilitated by a diverse range of technologies, two key ones being transportation, both of live animals and of animal products, and the preservation of food by refrigeration and freezing. Others are as diverse as electric fencing and genetically modified organisms, financial assistance in the form of direct and indirect subsidies, and agricultural research and extension services improving efficiency. The place of production is no longer determined by ecological factors but by a variety of interacting factors such as transport costs, the risk of disease, availability of labour, tax incentives and environmental impacts, as well as the supply of feed and the demand for products.

Fifthly, and related to technologies facilitating intensification, are social and infrastructural changes such as the organization of food supply chains. For example, vertical integration can provide economies of scale, reliability of supply, efficient management (keeping control of costs and product quality) and product consistency. Similarly, the emergence of multinational food processors and retailers changes the way food is grown, inspected, processed, packaged and supplied to consumers, often favouring preferred, usually larger, producers. For example, large industrialized pig farming operations in the US have costs that are a third less than traditional family farm operations. While smaller operations can stay in business, e.g. using family members as cheap labour, they tend to exit the industry as opportunities arise.

Finally, it is worth noting that not all systems can be industrialized, market incentives and farming as a way of life preventing the degree of industrialization seen in the poultry industry, for example. Dis-intensification can also occur as farmers abandon their land, in response to the same drivers of intensification (Van Vliet et al., 2015).

The accelerating agricultural intensification of the past half-century raises concern about whether farming is broadly ecologically sustainable, and especially whether it could remain so in future (MacLeod and Moller, 2006). Sustainability includes profitability and estimates of the increased costs producers would incur with free-range or other more humane systems of farming vary (Matheny and Leahy, 2007):

- a 0–3% increase for group housing sows;
- 1–2% for group housing calves;
- 5% for slow-growing broiler or meat chickens;
- 30% for free-range turkeys;
- 8–47% for free-range pigs;
- 8–28% for furnished layer hen cages;
- 8–24% for layer hens in barns; and
- 26–59% for free-range layer hens.

Table 3.2 *The performance and costs of production in free-range and more intensive systems. (1) Slow-growing chickens and standard housed meat or broiler chickens compared over three years (ITAVI, 2003). (2) Pig production modelled in free-range and fully-slatted, housed systems (data used were from a variety of scientific and commercial sources for four different stages of production and are thus presented as a range).*

1. Chicken meat production	Free-range	Intensive
Performance		
Stocking density (chickens/m^2)	11	22
(kg/m^2/year)	78	250
Flocks reared per year	3.3	6.2
Age at slaughter (days)	87	40
Live weight at end of rearing (kg)	2.2	1.9
Feed conversion efficiency	3.1	1.9
Mortality (%)	2.7	4.6
Production costs (€/kg liveweight)		
Fixed costs (depreciation etc.)	0.295	0.087
Operating costs (e.g. heating, veterinary charges, manure removal)	0.134	0.064
Feed	0.641	0.419
Chicks	0.126	0.121
Labour	0.222	0.037
Total costs of production	**1.418**	**0.729**

2. Pig production	Free-range	Fully-slatted, housed
Weaning to slaughter (days)	139	132
Average daily gain (kg)	0.45–0.73	0.38–0.86
Feed conversion efficiency	1.61–3.35	1.28–2.57
Mortality (%)	0.15–1.5	0.5–2
Production costs (p/kg carcass weight)		
Feed	59.9	49.7
Veterinary and medicines	0.9	1.4
Bedding	3.9	0
Housing	23.2	7.8
Labour	2.3	3
Energy	0	1.8
Water	0.7	0.8
Slurry/muck storage and disposal	0	1.8
Total costs of production	**99.3**	**94.7**

Table 3.2 gives a breakdown of the costs in different poultry and pig systems. While good welfare can pay, the relationship is curvilinear, so improving welfare in intensive or industrial-like systems also costs.

It is not just falling returns driving the need to intensify farming, but also greater expenses, greater expectations leading to increased costs, and a lesser share of the product prices. For example, fertilizer costs doubled, animal feed costs rose by 72% and energy by 66%, in the UK in the first five years of the 21st century. Farm working expenses have progressively risen over the last 50 years, from a 47% to a 61% proportion of NZ dairy farm income (Anonymous, 2014). Similarly, while wool prices have declined or remained steady, the cost of shearing sheep has increased, mainly due to increased labour costs, shearing accounting for over a third of the revenue earned from wool in 2005, compared with a seventh 20 years earlier (Anonymous, 2005). In some cases, for example, lambs that produce little wool, animals are often only shorn in order to manage the risk of flystrike, the income from the wool covering shearing expenses only and not the farm labour involved in mustering and presenting the sheep. Lambs' wool is a highly valued product used, for example, in cardigans, slippers and duvets.

The changing pattern is also seen in the smaller share of the value of food. For example, 50 years ago, British farmers received 50 pence out of every £1 spent on food; in 2005, it was just 7.5 pence (Benson, 2005). Between 1998 and 2006, UK farmers' share of the consumer price of untrimmed beef fell 20% and that of whole milk 9%, while that of eggs increased 2% (Anonymous, 2007). In the US, the farmers' share of the supermarket consumer dollar, 51 cents in 1918, had fallen to 24 cents by 1990 (Browne et al., 1992). In 2001, over 80% of the US food dollar went to value-added services and materials such as transportation, processing, distribution, labour, packaging and energy (Capps and Park, 2003). Or, expressed in perhaps a more understandable way, in 2000 the return to US farmers and ranchers from an $8 lunch consisting of barbequed beef on a bun, baked beans, potato salad, coleslaw, milk and a cookie, was just 39c (Lang and Heasman, 2004). This pattern is not limited to livestock farming; for every £1 spent by consumers on bananas, the Ecuadorian plantation worker received 1.5 pence, the plantation owner 10 pence, the trading company 31 pence, the ripener/distributor 17 pence and the retailer 40 pence (Lang and Heasman, 2004). It appears that the proportion that farmers receive for food has generally declined, the real profits of the last few decades having gone to the middlemen (Saul, 2005). Or as the poet Denis Glover (1912–1980) put it in *The Magpies* (1964):

Year in year out they worked
while the pines grew overhead
and Quardle oodle ardle wardle doodle
The magpies said

*But all the beautiful crops soon went
to the mortgage man instead
and Quardle oodle ardle wardle doodle
The magpies said*

Part of the cost-price squeeze is due to moves to add value to products. Instead of producing commodities, which others turn into high-value products for retailing, industries, regions and countries have turned to adding that value themselves. However, it seems to have created a more complex supply chain that captures the added value, leaving the farmers squeezed. It remains to be seen, or even appreciated, how much monitoring and quality assurance schemes will add to the costs of farming, further pressuring intensification.

These trends are not restricted to livestock farming and animal welfare but are also evident in crop farming (Fisher et al., 2009). A good or healthy soil grows or produces a good crop in local conditions, gives a good yield and is available for the next generation of farmers. However, cropping is one of the most intensive ways to degrade the soil, or at least requires greater inputs. Declining soil quality has been attributed to:

- more powerful cultivating machinery, which although more efficient, risks damaging or destroying the soil structure;
- economic or financial pressures, which have driven economies of scale, and changes from diverse systems towards monocultures with a loss of biodiversity and soil health;
- increasing competition for land use, which has forced cropping onto 'weaker' soils with increased risk of soil damage from cultivation and compaction; and
- fewer individuals with an intimate knowledge of the soil doing the work.

One of the main barriers, then, to looking after the soil is the cost: the financial pressure to farm profitably while doing so in a sustainable way. Sustainable practices can be costly, pressuring profitability. The cost-price squeeze is also evident in short-term ownership of land, especially that leased for cropping. Short-term leases, especially expensive ones, demand 'cash cropping' and do not encourage good practices. Less expensive longer-term leases provide the opportunity to better rotate crops and protect and enhance the soil. The reality of the cost-price squeeze is that while 'everyone likes what we're trying to do, nobody is prepared to pay'. There are no premiums for looking after the soil, no market advantage for those taking the lead in sustaining the soil.

Especially in difficult economic times, rising costs and low prices can result in poor animal welfare. In a review of farm animal welfare in the UK, the Food Ethics Council and Pickett (2014) stated:

There is no doubt that if you're looking at actual performance on the ground, it is often farmers that are strapped for cash, aren't employing enough people, aren't investing in decent kit, aren't investing in training, that are delivering the worst welfare on the ground. So that's not a standards issue – it's about taking the pressure off. Welfare suffers when farmers and the industry are under pressure.

Farming systems need to remain financially viable, especially in difficult economic circumstances when restricting animals' resources (e.g. space, bedding, staffing and labour, veterinary input) is the only option, resources that may be required to maintain animal welfare standards. Higher welfare can only be sustained if (1) consumers pay the additional costs of production associated with the higher welfare standards, (2) taxpayers bear the cost, for example, in the form of a subsidy covering the additional costs, or (3) the farm introduces efficiencies of production, for example, higher stocking densities, less farm labour or more productive animals. Such a change in thinking would require society to explore the cultural, social and political alternatives to farmers, processors, retailers, government and consumers benefiting from animal production. Should it be the almost exclusive expectation of farmers and their industries to absorb, to justify what many, if not all, members of society ultimately benefit from? Higher costs cannot be absorbed indefinitely unless higher returns accompany them. As farmers are under increasing economic pressure to increase the scale and efficiency of their operations to meet increasing costs and account for falling returns, intensification is inevitable. Alternatively, if there are societal limits to animal production, then how is society going to adapt when those limits are reached, the potential of biotechnology not withstanding? Many farming systems may be destined to progress towards a point characterized by questionable animal welfare, and economies of scale resulting in frequently unacknowledged environmental and social costs. As alluded to in Chapter 2, and captured in Figure 3.7, it is not just farmers, but many in modern society who are living off the back of farm animals.

The last part of this section is given to a summary of the Farm Animal Welfare Council *Opinion on the Welfare of the Dairy Cow* since it nicely captures the themes of this chapter, noting that some suffering arises from the poor profitability of dairying (Farm Animal Welfare Council, 2009). In concluding that the profitability of British dairy farming had fallen due to rising costs and lower prices for milk, the Council noted that although there were improvements, the welfare of dairy cows had not improved

Figure 3.7 Riding on the dairy cow's back (*Courtesy John Ditchburn*). (Available in colour in the plate section between pages 178–179.)

significantly. Amongst the evidence or rationale for this position was the belief that economic pressures were forcing farmers to seek greater efficiencies compromising cow welfare, low profitability reducing investment and maintenance. There was a 28% increase in milk yield per cow, a rise in sub-clinical mastitis, while the number of cows and dairy farms decreased. This was accompanied by a shortage of stockmen with farms heavily reliant on immigrant staff, and a decline in rural services. Profitability declined from an average gross margin of £933 per cow in 1997 to £696 in 2007, reflecting changes in exchange rates, the milk quota system, the price paid to farmers and greater exposure of the industry to commodity markets. The Council noted the demands of some retail buyers, as part of their supply contracts, to have lameness monitored. There was also an expectation that farmers be able to demonstrate how they have discharged their obligations to animal welfare legislation, and that public surveillance of cow welfare should be carried out to ensure progress with animal welfare was monitored and the findings given greater publicity. The Council then went on to recommend that the dairy industry should target a life span of at least eight years for their cows, and that breeding programmes should place more emphasis on animal welfare. Finally, the industry was encouraged to invest more in education, skills, training and professional development of farmers and stockmen, and on-farm recording, perhaps as part of farm assurance schemes, and it was recommended that health and welfare plans should be developed by the farmer and his or her veterinarian.

Although an internationally recognized and august animal welfare body, the Farm Animal Welfare Council was clearly, as Reynnells (2007) succinctly put it, introducing a symposium on the welfare and ethical concerns in poultry science: 'to manage a farmer's resources at no cost or risk to themselves'. It should be noted that the drivers of some of this low profitability were beyond the control of the farmer. To be fair, the Council is aware of those issues and has devoted resources to address them, and their role is in giving advice. They were also reflecting on developments in the dairy industry at the beginning of the 21st century. A decade later, there was oversupply and a significant fall in international milk prices borne of factors such as the removal of milk quotas in Europe, lower feed prices in the US, and trade sanctions resulting in fewer exports to some countries. These factors make farming difficult, especially when combined with societal concerns about animal welfare, the impacts on communities, the environment and food safety, and the distribution of wealth. In addition, these industries may well be suited to globalization, suiting countries with low costs, especially labour, and high corporate profits (Cheeke, 1999).

These examples affecting dairy cow welfare are revealing insights into what we will address in the next chapter, the expectations of how animals should be treated from those beyond farming. But I will end this chapter with two perspectives. The first is that of Alex Jack (2009), a NZ beef cattle farmer and Nuffield Farming Scholar who visited ten countries in 2009 to understand and respond to the multitude of factors (animal welfare, environmental and ethical) affecting farming. The second is a personal account by Richard Benson who in *The Farm* (2005) tells of the impacts on UK farmers and their families of a progressively global and faceless world drawing its wealth from agriculture.

Two reflections on modern high farming

Jack believed that if farmers cannot increase production, then they need to increase the return from or the value of their products to compensate for rising prices. Many farmers will not be able to afford new standards, as the cost of production will rise. This was quite sobering; there is no end in sight to the rising compliance costs and the costs of accessing the global marketplace. Farming must remain relevant in an era of dwindling political influence and in a rapidly evolving environment, by building understanding relationships with sympathetic groups such as Compassion in World Farming and the consumers they represent and influence without alienating consumers and voters. Farmer views have slipped out of step with the wider community; we need to consider how our practices look on YouTube or international TV without the benefit of explaining why and how they are undertaken. Furthermore, many farmers benefit from the pastoral perception of good animal welfare, even though they may make little conscious effort to earn it. Unless farmers show compassion, acknowledge those on the fringes who are prepared to push the limits for short-term financial gain, and seek help when they are not coping, society, the urban majority, may deem they are not worthy of the right to farm. Farm assurance is increasingly a part of modern farming; demonstrating good animal welfare and caring for the environment may not get a direct premium but is part of the cost of accessing affluent markets. The role the agricultural sector plays in sustaining the social and economic fabric of rural communities is not often appreciated until there is a crisis, for example, an outbreak of significant disease such as foot and mouth. Farming must continue to provide the marketplace with auditable, objective evidence of its role in protecting animals and the environment, supported by society and consumers with increased understanding and information about practices whilst being aware of their sentiments and concerns.

Benson (2005) invites the reader to begin by looking at an English village in the 1980s where a mixed farmer and contractor finds it difficult to stay profitable in the face of increasingly volatile interest rates. He had taken advice to specialize in pig farming utilizing modern science and technology, but that helped produce more food, resulting in a glut in the marketplace and lower prices. Furthermore, corporate retailers replaced once specialized grocers and butchers, making it more difficult for individual farmers to supply individual retailers, while food processors demanded consistent and increasingly cheaper meat, often without regard for consumer wants. Farm subsidies favoured big producers and the mantra of 'more food, more efficiently, more cheaply' contributed to the loss of smaller farms and businesses. As the costs of production rose, more food was imported, and more farmers had to sell their assets to remain solvent:

> *If you listen carefully, you can hear a few, a very few people, suggesting that the rush to gain cheaper food is leading to the loss of something which might be officially out of date but somehow seems a good thing that it seems wrong to lose. The British mixed farm was a creation of the enclosures in the eighteenth centuries and had by the end of the 1990s all but disappeared. Its point was to combine livestock and arable so that a) if one crop failed you had others as back-up, and b) there was an interlinking of production in that you could feed crops to the animals, and fertilize*

the land with animal manure. To many people unfamiliar with agriculture, this seems so perfect and commonsensical an idea that they are vaguely surprised to find out that it is now an eccentricity, a defiance, a tourist-centred enterprise. For some of the few people who murmured about something good being lost, this is one of the cases of money defying common sense that encapsulates the jarring, perplexing fact of the modern world: that the logic and geography of business is not syncopated with the logic of human feelings.

The most intensive or industrial farming systems producing dairy, poultry and pork are common because they are commercially successful. Their meat, milk and eggs are sold in supermarkets frequented by relatively affluent people, making up a significant proportion of these people's protein intake. But they have also attracted, especially in modern affluent societies, the severest criticism because of their impacts on animal welfare. As it is not just the farmer who benefits from increasingly more expensive and highly productive farming, we might also ask, who guides, influences and audits the expectations of consumers and others in society? It is to the history and extent of this criticism that we now turn.

Further reading

The Agricultural Revolution in Prehistory. Why Did Foragers Become Farmers? **Graeme Barker**. Oxford University Press, Oxford, UK, 2006.

The Farm. The Story of One Family and the English Countryside. **Richard Benson**. Penguin, London, UK, 2005.

Growing for Good: Intensive Farming, Sustainability and New Zealand's Environment. **Parliamentary Commissioner for the Environment**. Wellington, NZ, 2004.

The Spirit of the Soil. Agriculture and Environmental Ethics. **Paul Thompson**. Routledge, London, UK, 1995.

Agriculture's Ethical Horizon. **Robert Zimdahl**. Academic Press, Burlington, MA, 2006.

CHAPTER FOUR

Husbandry from beyond the farm gate

The small town of Bagshot in southern England, part of the turnpike between London and the west, has a long history dating back to at least the Bronze Age and Roman times. It is here also that the 19th horse was provided for Captain Lapenotiere's 271-mile, 37-hour ride, using 21 horses, from Falmouth to London to inform the Admiralty, and ultimately the Prime Minister and the King, of the defeat of the combined French and Spanish fleet at the Battle of Trafalgar in 1805, and of the death of Vice Admiral Lord Nelson. The town's more notorious but nonetheless interesting historical link is Jolly Farmer Corner, a reference to a pub, originally named the Golden Farmer, supposedly after a local highwayman whose wife paid their debts in gold. While a convicted highwayman, John Bennett, was executed in 1689, and his body gibbeted at Bagshot Heath, it is likely that the story has become entwined with that of a local farmer William Davis, his wife and their 18 children. However, it is the history on the pavement of one of Bagshot's streets, an old trough full of flowering plants (Figure 4.1) inscribed with 'Be kind to all God's creatures', that is most relevant. It is one of many remnants of an exceptional undertaking begun in the 19th century by The Metropolitan Drinking Fountain and Cattle Trough Association to provide clean and free water for people and animals, especially carriage horses and driven livestock (Anonymous, undated, a). At a time when farm animals were walked to market and slaughter, horse-drawn vehicles packed streets and palatable water was unobtainable, pure and cold water was provided for man and animals with many troughs installed in London and throughout England. While some are still in use, many are now maintained as flower-planters.

Significantly, the provision of water was a practical means of caring for animals. More recently, a consortium of animal welfare groups, including the Humane Society of the United States and the American Society for the Prevention of Cruelty to Animals, helped fund the development of a

Figure 4.1 Drinking troughs provided by the Metropolitan Drinking Fountain and Cattle Trough Association from the latter half of the 19th century (*Photo: Alamy*) and more recently (*Photo: Mark Fisher*).

centre-track conveyor restrainer system for the humane handling of animals at slaughter plants (Grandin, 2003). There are no doubt many other examples of individuals and groups quietly working to improve the lives of farm and other animals. However, there is also no shortage of advice and information on how farmers should look after their animals, perspectives that may or may not be adequately informed, as in Benson's (2005) reflection: 'In the end, I was just another city person imposing a set of ideas on the countryside that the countryside had never claimed for itself.' Furthermore, and unlike the provision of drinking troughs, the demands that animals are treated in certain ways often come with little cost or risk to those demanding change, as Reynnells (2007) noted.

Although animal welfare is essentially about the state of an animal or what it feels, what we understand about the significance of what we think an animal might feel, has many dimensions. We all have views, rightly, on how animals should be farmed, transported and slaughtered and many invest significant resources in expressing those views and demanding changes to practices. What we often fail to do, however, is understand those practices in the context of others' views, to see the big picture, the historical, scientific, ethical, economic, political and pragmatic dimensions of animal welfare. For instance, the complexities and difficulties are revealed in the reasons modern farmers gave for not implementing advice emanating from an animal welfare assessment study. They included a lack of resources (financial and time), practical difficulties of implementing changes that would also lead to other problems, or the view that current systems just did not need improving. Similarly, the belief amongst those conducting the study, that farmers could conduct part of a formal animal welfare assessment, thereby saving time and costs, overlooks the farmers' time and costs (Botreau et al., 2013).

The characteristic human trait of having and voicing an opinion, often from different understandings and perspectives, and often without understanding the context, has spawned a diverse, influential and important 'animal welfare industry'. There is a long history of such views, beginning in values, myth and legend, and more recently in the literature, both scientific and general. However, it is well to remember that such views are recorded not only because they deal with the tension between animals and humans, nature and culture or sentiment and reality, but also because they differ from, and challenge, the accepted views of communities and societies. Like any history, they reflect the perspectives of the writer, not necessarily those interacting with animals.

Like the archaeological layers excavated below settlements and cities, modern civilization is built on a legacy of values (Comstock, 2000), much of it, as hunter-gatherers, beyond the understanding of modern western people (Nelson, 1993). Hunter-gatherers were guided by myths and legends, narratives that depicted animals as having spiritual powers. The principles supporting hunting included beliefs that animals have a spiritual consciousness and individual identity requiring respecting the dignity of animals by adherence to rituals, practices acting to restrain human use of animals (while entitled to hunt them, humans must exist in balance with animals). As humans took up farming, they began to believe that animals were made by God for humans, but that those animals had to be respected and had needs which must be provided for. The husbandry ethic of the shepherd guiding and caring for the flock, so important it is more often associated with pastoral care of humans, is part of our expectations for stockmanship and the humane care of animals. So it is to have a good life and, when the time comes, a quick and painless death. The interdependence of farm animals and humans is also central to the husbandry ethic – it is appropriate to use animals for human benefit if that use is humane (in modern terms, justified and any harms minimized, Banner et al., 1995). However, there are a wealth of views, commandments, rules, laws, standards, guidelines, directives and regulations, outlining both expectations of what humane care involves, and challenges to the predominant view of using animals. The following is not a comprehensive summary but an eclectic mix of some of the more interesting, obscure and influential.

The *Book of the Dead* to the era of *Animal Machines*

In the 16th century BC, the Egyptian *Book of the Dead* (or *Spells for Going Forth by Day*), a collection of spells to guide the soul of the deceased through the afterlife, warned that ill-treatment of animals harms the passage of the dead to eternal life (Preece, 1999). The spells gave power to the deceased, keeping them safe and allowing them to turn into different animals to make it easier to travel in the afterlife. There are many other early recorded attitudes to animals. For example, a somewhat similar belief was held by the Greek philosopher Pythagoras (c.570–490 BC) who regarded all living things as kindred, their souls transmigrating to other animals after death. Pythagoras held that politicians needed to show justice towards animals and refused to eat meat, the 'Pythagorean Diet' (influenced by the Ancient Egyptians) preceding the term vegetarianism by 5000 years (Preece, 2002; Rauw, 2015). And in the 3rd century BC, Asoka (304–232 BC), a King in northern India, abolished sacrificial slaughter, banned the royal hunt and planted trees to provide shelter for animals as well as people (Knierim and Jackson, 1997).

The relationship between humans and animals and the treatment of the latter are addressed in many traditional religions. The Islamic tradition strongly supports humane slaughter: 'Verily Allah has enjoined goodness to everything; so when you kill, kill in a good way and when you slaughter, slaughter in a good way. So every one of you should sharpen his knife, and let the slaughtered animal die comfortably' (Rahman, 2017). While care of animals is part of the pastoral ethic in the Bible, there are also calls for compassion. For example, 'If you see the donkey of someone who hates you fall under its load, do not stand back; you must go and help him with it' (Exodus 23:5). Similarly, from the Talmud, 'Rejoicing cannot occur at an animal's expense' and, 'Jews must avoid plucking feathers from live geese, because it is cruel to do so.' The relationship between, and similarity of, animals and humans are also stressed, as in the Bible's 'for the fate of humans and the fate of animals is the same; as one dies so does the other' (Ecclesiastes 3:19). Finally, there is the Hindu observation that, 'Everyone in the [meat] business, the one who cuts, the one who kills, the one who sells, the one who prepares, the one who offers, the one who eats; all are killers' (Preece, 2002).

In antiquity, Lucius Junius Moderatus Columella, a Roman soldier and farmer (4–70 AD) who wrote extensively on farming, suggested that hens should be able to dust bathe following the observation by Heraclitus (c.500 BC) 'that pigs wash in mud and farmyard birds in dust or ashes'. Born in Spain, Columella moved early in life to Italy where he owned farms and lived near Rome, after military service in Syria and Cilicia. Columella's *On Agriculture* (*De Re Rustica*) is a comprehensive, systematic and detailed account of Roman agriculture (Columella, undated). Various books cover the farming site, water supply, buildings and staff; ploughing; fertilizing, and care of crops; cultivation, grafting and pruning of fruit trees, vines and olives; the acquisition, breeding and rearing of oxen, horses and mules, and veterinary medicine; sheep, goats, pigs and dogs; poultry and fish ponds; bee-keeping; gardening; the duties of the overseer and the overseer's wife; pickling; and preserving and making wine. Another work dealt with vines, olives and various trees. Some of Columella's advice on husbandry would not be out of place in animal welfare today. For example, hens should be provided with perches to prevent their dung harming their feet; the poultry-yard should be free of moisture; dry dust and ash should be provided for dust bathing; housed birds should have access to a spacious porch-like area, protected from eagles or hawks, in which they can bask in the sun. However, Columella also noted, regarding young pigeons, that:

> *It is only worth while to go to these expenses and to take these precautions in places where the Young pigeons indeed are more easily fattened under their mothers' care, if when they are already strong but before they begin to fly, you pull out a few of their wing-feathers and crush their legs, that they may remain quiet in one spot, and give plenty of food to the parent-birds with which they may feed themselves and their young more abundantly. Some people bind their legs loosely together, because they think that if they are broken, pain, and consequently emaciation, is caused; but doing so does not contribute at all to their fattening, for, while they are trying to get rid of their bonds, they are never at rest, and by this kind of exercise, as it were, they add nothing to their bulk. Broken legs cause pain for not more than two or at most three days and deprive them of all hope of wandering abroad.*

Figures such as René Descartes (1596–1650), a French philosopher, mathematician and scientist, are often portrayed as being influential. Descartes (1976) held that bodies were like machines, made by the hands of God and incomparably better arranged than machines made by man, and like machines, both humans and animals could walk and eat without thinking or reasoning. Descartes declared: '… all the things which dogs, horses and monkeys are made to do, are merely expressions of their fear, their hope, or their joy; and consequently they can do these things without any thought'. He used speech and the fact that machines did not act with the benefit of knowledge, to differentiate men and brutes. As humans require very little reason to be able to talk, animals, because they do not have speech, cannot reason. It is nature that allows them to act, just like a clock. Furthermore, as animals express their passions – fears, hopes and joys – they would express their thoughts if they had any, speech being the only certain sign of thought. Interestingly, Descartes, who was discussing thoughts and not life or sensation that he did not deny in animals, commented that his position was not so much cruel to animals as indulgent to men, absolving them from 'suspicion of crime' for killing and eating animals.

Many have interpreted Descartes' *bete-machine* as a monstrous thesis, that animals are without feeling or awareness of any kind. Others have held that there is no evidence that Descartes believed that animals had no consciousness or that they were totally without feeling (e.g. Cottingham, 1978). Whether history has been true to Descartes or not, it is doubtful that a 'monstrous' view would have displaced common sense and his views were not shared. For example, Voltaire (1694–1778), the French writer, philosopher and historian, noted that we judge other humans without resorting to their speech: for example, appearing disconsolate and anxious looking for a paper, then joyful at finding and reading it. Similarly, we judge a dog's feelings of distress and pleasure on losing and then reuniting with its master. David Hume (1711–1776), the Scottish philosopher, historian and economist, stated: 'Next to the ridicule of denying an evident truth, is that of taking much pains to defend it; and no truth appears to me more evident, than the beasts are endow'd with thought and reason as well as men.' He held that when animals act like humans for similar reasons, then it is reasonable to believe in like causes. Finally, Johann Wolfgang von Goethe (1749–1832), a German writer and scientist, held that machines could be dismantled and assembled again but a living organism could not as the whole shaped the parts whereas in machines, the parts shaped the whole. Perhaps Descartes' *bete-machine* is better known, or at least referred to, for its monstrousness than for any real belief in animals acting like clocks.

The Renaissance brought a belief that men were the only worthy beings, though animals were regarded as sentient and intelligent. However, the term cruelty, referring to the evil intentions of foreign or religious groups or to the nature of beasts, began to be used to describe the killing of animals. In 1635, as part of 'civilizing' and 'anglicizing' Ireland, it was prohibited to pull the wool off sheep, or attach a plough to a horse's tail. By 1641, the New England colonists in North America decreed that, 'No man shall exercise any tyranny or cruelty to ward any brute creature which are usually kept for man's use' (Ryder, 2000).

The 17th century also saw the early sentiments of the modern movement for the ethical treatment of animals, the Englishman Thomas Tryon (1634–1703) campaigning for human responsibilities to animals. At the age of 12, Tryon was made to work at his father's employment after which he became a shepherd. With the sum of three pounds from the sale of his four sheep, at 17 Tryon went to London to seek his fortune. He began writing at 48, and penned, in Victorian style, the verbosely entitled *The Country-Man's Companion: OR, A New Method Of Ordering Horses & Sheep So as to preserve them both From Diseases and Causalties, Or, To Recover them if fallen Ill, And also to render them much more Serviceable and Useful to their Owners, than has yet been discovered, known or practised. And particularly to preserve Sheep from that Monsterous,*

Mortifying Distemper, The Rot (Tryon, undated). Tryon gave the following view of the world from the farm animals' perspective:

> *We COWS give him our pleasant Milk; which is not only a most Sovereign Food of it self, but being altered, and variously dressed, makes a great number of delicate and wholesome Dishes; but this will not content them; for after we have for several Years twice a day yielded them plentiful Meals of Milk, and every Year (for the most part) a Calf; which they have rended away from us before its time, to our great affliction: (for 'tis unnatural to take away any Creature's Young, before it can provide for it self) when they have thus bereaved us of the Fruit of our Wombs, and killed, and eaten them, yet sometimes, through Covetousness, notwithstanding all these Benefits, they will half starve us; and if it chance that any one of us do not give good store of Milk, and bring them forth as many Calves as their Avarice expects, then our Egyptian Masters cry, Hang her, knock her o'th' Head; what is she good for?*

The German philosopher Immanuel Kant (1724–1804) held that animals were sentient though without reason or understanding. Cruelty to animals was considered wrong not because of the harm to the animal, but because it might result in humans becoming cruel to each other. Interestingly, the need to protect society from the ill-disciplined spectators, rather than concern for animals, drove the early reform of practices such as bear-baiting and cock-fighting. Jeremy Bentham's famous dictum 'The question is not, Can they reason? Nor Can they talk? But Can they suffer?' needs little explanation. The statement, apparently made in an 'obscure footnote' (Preece, 1999), is now central to much of animal welfare; exploiting animals was unacceptable because they were sentient. Bentham (1748–1832), an English philosopher, also stated that rights were 'nonsense on stilts' and that depriving an animal of its life was justifiable since 'their pains do not equal our enjoyment'.

The 19th century has been described as the legislative era (Preece, 2002). Thomas Erskine (1750–1823), a lawyer and politician who supported the emancipation of slaves and was instrumental in developing laws on behalf of animals, stated:

> *... we should never forget that the animal over which we exercise our power has all the organs which render it susceptible of pleasure and of pain. It sees, it hears, it smells, it tastes. It feels with acuteness. How mercifully, then, ought we to exercise the dominion entrusted to our care!*

Fourteen anti-cruelty bills were introduced into the British Parliament, most of which were defeated. Those which were eventually enacted related to cruelty to horses and cattle; bull, bear and badger baiting and cock and dog fighting; cruelty to dogs; and some blood sports. Erskine was a driving force behind what became known as Martin's Act, Richard Martin being responsible for the passing of the first national anti-animal cruelty law, *An Act to Prevent the Cruel and Improper Treatment of Cattle*, in England in 1822.

Henry Salt (1851–1939) was an English writer and campaigner for social reform. He prefaced his 1892 *Animal Rights: Considered in Relation to Social Progress* by noting that many of his 'contentions will appear very ridiculous to those who view the subject from a contrary standpoint, and regard the lower animals as created solely for the pleasure and advantage of man'. Salt held that animals, as well as men, though to a lesser extent than men, possess distinctive individualities and are thus free to do what they want provided they don't infringe upon the equal liberty of others. In other words, if humans had rights then so did animals. Drawing on a number of moral works, Salt considered the antiquated 'great gulf' between animals and humans, dismissing the reasoning such as the religious notion that animals 'have no souls' and thus no prospect of an afterlife to compensate for suffering in the present one (unlike a church dignitary of the time who allowed vermin to bite him without hindrance since the 'poor creatures have nothing but the enjoyment of this present life'). Salt noted the growing belief that mankind and animals have the same destiny before them 'be it immortality or annihilation'. Animals, then, had a right to live a natural life, a life that permits the individual to develop subject to the limitations of the needs and interests of the community. Salt then went on to state:

> *If we must kill, whether it be a man or an animal, let us kill and have done with it; if we must inflict pain, let us do what is inevitable, without hypocrisy, or evasion, or cant. But (here is the cardinal point) let us first be assured that it is necessary; let us not wantonly trade on the needless miseries of other beings ...*

The difficulty of determining the real or necessary requirements of the general community was acknowledged: 'no words can be devised for the expression of rights, human or animal, which is not liable to some sort of evasion, and all that can be done is to fix the responsibility of deciding what is necessary and unnecessary, between factitious personal wants and genuine social demands ...' This was trusted to the interaction of the personal conscience of individuals and the public conscience of the nation. The 'divorce of humanism from humaneness', noted Salt, 'is one of the subtlest dangers by which society is beset': the disconnection of man from nature, the result of our persistent reliance on reason and neglect of the emotional, the spiritual or the instinctive, the price paid for civilization.

Albert Schweitzer (1875–1965), a French theologian, philosopher, physician and musician who became a medical missionary in French Equatorial Africa, considered reverence for life, all life, as the basis of ethics (Schweitzer, 1936, 1987). Rejecting Descartes' 'I think therefore I am' as the starting point for philosophy, Schweitzer noted that all organisms have a will to live and that this should be the basis of our thinking ('I am life which wills to live, in the midst of live which wills to live'). His reverence for life ethic, for which he was awarded a Nobel Prize, held that: 'it is good to maintain and cherish life: it is evil to destroy life and to check life.' However, Schweitzer was a realist, noting there were times when life had to be destroyed, but only when necessary and after careful reflection ('for a higher value and purpose – not merely in selfish or thoughtless actions'), even in seemingly insignificant cases. It is something for which to strive, telling us 'that we are responsible for the lives about us'. Schweitzer also held that, 'Wherever any animal is forced into the service of man, the sufferings it has to bear on that account are the concern of every one of us.' We are all guilty and must bear the blame for animal maltreatment, 'the cries of thirsting creatures', the 'roughness in our slaughterhouses', or at the hands of 'the dreadful play of children', forbidding us to cease to feel what animals endure. As humans are interconnected and interdependent with the living world, moral responsibility extends to all living things, ethics being a 'responsibility without limit towards all that lives'. Schweitzer believed that the disconnection between material and spiritual development was at the basis of the disconnect, contributing to the decline of the western world. It was therefore important to have a direct connection with the living world and only then would we be able to take responsibility.

However, it was Ruth Harrison's *Animal Machines* (1964), a devastating account of the conditions calves, pigs and hens experienced in the early 1960s, that brought animal welfare to the attention of the wider community. A *Crusade Against All Cruelty to Animals* leaflet describing how animals were raised for food provided the inspiration for Harrison (1920–2000) to investigate factory farming. A vegetarian, she was driven by a sense of injustice done to animals that deserved to live a life, rather than merely exist, before they were slaughtered. Extensively informed of veterinary, farming and scientific material, and well-reasoned, *Animal Machines* highlighted that animals could suffer distress and discomfort despite thriving, in contrast to farmers insisting animals would not grow and produce if they were unhappy. If animals suffered, it was held, so would production. Harrison argued that to ensure the well-being of animals, state intervention was required. Harrison was not the first to highlight the problems of factory farming, but it was the novelty of her message, her sincerity and resonance with the impacts of industrialization on society, and her reliance on information and informed engagement that meant her critique was devastating (Harrison, 1964, 1988; van de Weerd and Sandilands, 2008; Woods, 2011b).

In response to Harrison's book, which was serialized in the *Observer*, a national UK newspaper, the Royal Society for the Prevention of Cruelty to Animals (RSPCA)'s lobbying and the public outcry, the British government commissioned an examination of the conditions of intensive husbandry livestock. Chaired by Professor F.W. Rogers Brambell, the *Report of the Technical Committee to Enquire into the Welfare of Animals Kept under Intensive Livestock Husbandry Systems* (commonly the Brambell Report, 1965) concluded that intensive husbandry methods should not necessarily be regarded as objectionable and may often benefit animals. However, certain practices were contrary to animal welfare and needed to be controlled. For example, the Report stated that the close tethering of calves, debeaking of some poultry and routine tail docking of pigs should be prohibited. In addition, the Committee recommended that guidance on numerous good husbandry practices be incorporated into advice provided to producers and acknowledged the importance to animal welfare of the qualities and skills of those responsible for animals. Noting that domestication necessarily involves some measure of confinement, the degree of confinement or restraint characteristic of all intensive systems was seen to unnecessarily frustrate natural behaviour. It was held that an animal should at least have sufficient freedom of movement to be able without difficulty, to turn around, groom itself, get up, lie down and stretch its limbs (the original five freedoms). The Committee rejected the idea that productivity was a decisive measure of welfare, regarding health and the ability to behave naturally as also important.

Animal Machines, the Brambell Report and public debate led to a 1968 Act addressing the welfare of livestock, the formation of the Farm Animal Welfare Council (to advise the Government), codes of welfare, a place for farm inspections, enforcement and advice, and, overall, animal welfare's place in the public domain or psyche today. Interestingly, these events brought a change from the view that cruelty and suffering were only evident if animals did not thrive in a farming productive sense, to one in which emotions and natural behaviour could also be considered. Scientific knowledge (often lacking) and practical experience (compromised by economic interests) did not provide an adequate basis for determining how animals should be kept; their well-being was a moral issue rooted in human attitudes. It led to the articulation that while the relationship between animal well-being and profit could be determined by scientific and practical knowledge, determining the balance between them was an ethical decision (Harrison, 1988). The productivity and feelings views of animal well-being, for a number of reasons, resulted in the term animal welfare being used in preference to cruelty and suffering, perhaps to appease the farming community and reduce public criticisms of it, perhaps reflecting the growing importance of welfare in other parts of society. Whatever political expediencies it served, the moral and ethical dimensions of animal welfare were acknowledged, the tension between empirical and social perspectives a reflection of the complexity and contentiousness of the issue then, as it is now. How animals were to be treated became a matter of ethics informed by science and other forms of knowledge. Harrison concluded that the 'arguments against factory farming are essentially based on humanitarianism and quality: the arguments for factory farming, such as they are, are economic arguments', something that she noted we should not be disturbed about.

Whereas suffering or cruelty was described by William Thorpe in the Brambell Report as the suppression of the basic urges of feeding, locomotion, grooming, social responses, sleeping and rest, it was redefined as 'unnecessary pain or unnecessary distress' in the 1968 law (Harrison, 1971). In debate about the value of codes of welfare, it was noted that: 'the debate is not about animals; it is about human beings and about the way in which they are in a civilized society treating animals in that society.' Similarly, it was pointed out that: 'if you look at economic factors without reference to ethical ones ... there is a backlash which is damaging not only to nature and to man but also, ultimately to the economic advantage which is sought.' In contrast, some classic statements were attributed to the other side: 'Turning around may not be advisable from the point of view of the animal' and 'I doubt whether there are many children in this country who live in such comfortable circumstances [three-bird battery cage] as some of these birds do ...' (Harrison, 1971). Marian Stamp Dawkins, a British ethologist with research interests in animal consciousness and animal welfare, and author of *Animal Suffering: The Science of Animal Welfare*, noted in a presentation to the 2014 International Conference on the Assessment of Animal Welfare at Farm and Group Level that there are lessons still to be learned from *Animal Machines*. For example, for food safety, Harrison stating: 'Unhealthy animals cannot make healthy food for humans'; animal health: 'Conditions are so crowded that any disease can sweep through the house very rapidly'; and even economics: 'One begins to wonder whether it might not be cheaper to eliminate the labour shortage in this field by raising the status of the agricultural worker.'

Imagine standing at a fourth-story window, tempting passerby's with a free-meal, only to hook them, haul them up and batter them to death 'without risk of reprisals or recriminations'. So began an article in the UK's

Sunday Times in 1965, the same year the Brambell Report was published, by Brigid Brophy (1929–1995), a novelist and short-story writer (Brophy, 1966; also see Brophy, 1971). She was once described as having 'more brains than anyone deserves' and 'a brilliant stylist' (for example, 'The bullfighter who torments a bull to death and then castrates it of an ear has neither proved nor increased his own virility; he has merely demonstrated that he is a butcher with balletic tendencies'). As well as the rights of animals, Brophy also campaigned for, among other things, the rights of authors to be paid when their books were borrowed from libraries. *The Rights of Animals* went on to describe the relationship between humans and animals as 'unremitting exploitation', eating, wearing, serving our superstitions, exhibiting and taking pleasure in their killing. Brophy then noted that whereas we contest where the rights of one man end and those of the next man begin, 'only in relation to the next animal can civilised humans persuade themselves that they have absolute and arbitrary rights – that they may do anything whatever that they can get away with.' Brophy's argument was based on animals being sentient and that, therefore, we have 'a moral obligation to spare them pain and terror' in rearing and killing them for our food. The right to live is independent of the right to shun pain, and personally, she believed that even if we did not cause pain, we don't have the right to kill animals any more than we have the right to kill humans. Throughout the article, Brophy explores a reader considering her a crank, a killjoy, even an anthropomorphist, but provides eloquent insights questioning those conclusions. She maintained that animals are not equal to humans but that, being capable of imagination, rational and moral choice, humans are superior and thus obliged 'to recognise and respect the rights of animals.'

Another part of this early modern era was Richard Ryder's concept of speciesism, initially outlined in a leaflet circulated around Oxford. Noting that there were no essential or unique differences between humans and other animals, Ryder (1970) asked why humans should make a moral distinction. If it is wrong to inflict suffering upon humans, then that right should be extended to animals as well. Like race and racism, or sex and sexism, species alone was not cause for inflicting suffering (Ryder, 2000).

The popular understanding of 'animal rights' in modern times was subsequently advanced by two very influential books, *Animal Liberation* by Peter Singer (1975) and *The Case for Animal Rights* by Tom Regan (1938–2017). Singer advanced the principle of equal consideration of interests: whereas it is bad to inflict pain on a human for no good reason, it is equally bad to inflict it on an animal for no good reason. Regan (1983), in *The Case for Animal Rights*, put forward the view that animals, in having, for example, desires, memory, emotions, preferences and a sense of what fares well or ill for them, are the subjects-of-a-life and as such have a fundamental right to be treated with respect. For Regan, this 'right to life' means many forms of human–animal interactions, including commercial agriculture and animal research, are intolerable, a position he notes would have profound, even revolutionary consequences. Humans are in relationships with animals, as Mary Midgley (1983) contends in *Animals and Why They Matter*. Few people see the social world as exclusively human and our capacity for extending sympathy beyond the familiar group is an important part of being human. It is these relationships, what Midgley terms as being part of a mixed community, which many formal ethical theories fail to acknowledge. Just as we see our families as different or special in some ways, what we decide is right and wrong must flow from these important relationships (Anderson, 2004).

Experts, such as philosophers, traditionally have an important role in challenging society's acceptance of, or attitudes to, subjects such as animal welfare. However, it is important to be aware that they are influential because they present views that challenge traditional values, those who defend some current practices less well known. Care should be taken in discerning common attitudes towards animals from the deliberations of philosophers, especially where, as is the trend in modern times, these are often used in advocacy rather than in understanding.

Expressions of modern humanitarian concerns

Despite humanity's long concern with the treatment of animals, animal welfare has entered the public domain with many believing 'animal welfare is becoming increasingly important'. The public is an important driver of animal welfare since, in democratic societies, compromises to animals are deemed necessary and reasonable by consensus. There have been many relevant events and a variety of sociocultural factors contributing to that public position. One of the insights of history is that the treatment of animals has moved from seemingly being mainly the responsibility of individuals engaging with animals to having greater explicit societal oversight, for example, being encoded in society's social arrangements such as the law. According to Bernard Rollin (1993), a philosopher at Colorado State University, it was a response to modern science and technology (antibiotics, vaccines, genetics etc.) facilitating greater efficiency and productivity, resulting in cheap and plentiful food. Farming was no longer constrained by an animal's biology or the more traditional mores of avoiding pain and distress and providing conditions to which animals could adapt; technology 'allowed us to divorce animal productivity from animal happiness'. While not trying to cause cruelty in the intentional or malicious sense, suffering was the result of lack of space and companionship, an inability to move freely, and boredom. Furthermore, the scale of those operations, and the margins received, both prevent normal husbandry procedures such as the detection and treatment of individual animals experiencing suffering and make it unaffordable to treat them. Rollin holds that society has rejected the values of efficiency and productivity associated with intensive confinement systems and upheld the husbandry values of traditional farming that allows animals to live according to their natures. This new or emerging social ethic, which arguably began with the publication of *Animal Machines* by Ruth Harrison (1964), reinforced by the recommendations of the Brambell Committee, highlighted the importance of considering needs that do not necessarily impact on efficiency and productivity, or at least can be masked by technology, namely social needs such as companionship and relief from boredom. While some farmers may value productivity, a measure of a system's economic success, it is argued that society values all dimensions of animal life: not just their physical health, growth and reproduction, but also animals' mental health and opportunities for positive experiences. Most social decisions are made to benefit the larger group, but not necessarily at the expense of the fundamental interests or rights of individuals. Thus, just as society extended its ethical concerns to include women, different races and minority groups, it has extended them to animals, constraining people from realizing the benefits of efficiency and productivity without addressing the concerns of other dimensions of animal well-being (that animals do not suffer and that their needs are met according to their natures). It requires that the values of efficiency and productivity are balanced with the values of husbandry and well-being, demanding that environments are designed to satisfy animals' needs.

This new or emerging social ethic is expressed and addressed in many ways, reflecting the range of responses that animals evoke, from the ideological to the practical and the empirical to the sentimental. Consequently, there is a complex and at times confusing array of evidence used to support various positions. This evidence can be drawn from science, anecdote, tradition, history, philosophy, values and ethics, personal beliefs, hearsay, experts, professional advice, and the interpretation of legislation, economics, political and market impacts. While each has its merits, insights and limitations, they are also favoured or rejected because they reflect and/or represent world views and decisions about what is and is not acceptable. Animal welfare policy must draw on, or at least acknowledge, all. Results and observations, beliefs and opinions are interpreted by humans within the circumstances or context they find themselves in, using not only reason but also common sense, ethics, imagination, intuition and memory (Saul, 2001). Or, as one reflection of modern politics suggested, 'Expertise can only get you part of the way. More valuable by far are experience, wisdom, independent judgement – and common sense' (Norman, 2013). Sources of insights can include popular works like *Black Beauty* and *Animal Machines* that have drawn attention to animals, as have works like Upton Sinclair's (1878–1968) *The Jungle*, which set out to highlight employment conditions at meat processing plants but also drew attention to food safety and animals. *Animal Machines* led to the establishment of the UK's Farm Animal Welfare Council, now the Farm Animal Welfare Committee. The Committee provides advice on the welfare of farm animals on agricultural land, at market, in transit and at the place of slaughter. Other countries have similar bodies such as NZ's National Animal Welfare Advisory Committee, and one of the earliest, the Animal Welfare Board of India, was established in 1962.

Evolving from what became known as the 'five freedoms' of movement identified in the Brambell Report, under the tutelage of John Webster and the Farm Animal Welfare Council, the Five Freedoms and Provisions were articulated (Webster, 2005a):

(1) *Freedom from thirst, hunger and malnutrition* – by ready access to fresh water and a diet to maintain full health and vigour.
(2) *Freedom from discomfort* – by providing a suitable environment including shelter and a comfortable resting area.
(3) *Freedom from pain, injury and disease* – by prevention or rapid diagnosis and treatment.
(4) *Freedom from fear and distress* – by ensuring conditions that avoid mental suffering.
(5) *Freedom to express normal behaviour* – by providing sufficient space, proper facilities and company of the animal's own kind.

The Five Freedoms have become very influential and are probably the most singularly recalled or identified feature of contemporary animal welfare, paraphrased as the needs of animals in some countries' legislation. The Five Freedoms have been reformulated as the Five Domains, understanding the needs of animals in terms of compromises and opportunities for nutrition, environment, health, behaviour and mental state, encapsulating both positive and negative states of welfare (Mellor and Beausoleil, 2015; Mellor and Reid, 1994). Similarly, there is the concept of an animal's Quality of Life: a good life, a life worth living, a life worth avoiding and a life not worth living (Fraser, 1993; Green and Mellor, 2011; McMillan,

2000; Yeates, 2011). Although grounded in science, these frameworks are not distinct from values, providing the motivation or aspiration to enhance, or at least maintain, animal welfare.

There have been other animal welfare-related government enquiries and reports and one of the more erudite was that chaired by Reverend Professor Michael Banner, producing The *Report of the Committee to Consider the Ethical Implications of Emerging Technologies in the Breeding of Farm Animals* (Banner et al., 1995). Although the Committee's work related to breeding and animal welfare, it began by articulating the principles guiding the use of animals. Reflecting a long history of thought about the relationship between man and animals, the common view was that the use of animals is acceptable providing it is humane. However, the Committee held that some harms should not be inflicted upon animals, that any harm should be justified, the benefits must outweigh the harms to the animal, and any harm so justified should be minimized as far as reasonably possible. Bodies such as Banner's typically provide opportunities for representation from interested parties with many individuals and groups contributing or making submissions.

Internationally, the importance of the various dimensions of animal welfare is reflected in the World Organisation for Animal Health, better known by its acronym OIE (Office International des Épizooties), view that animal welfare is a complex, multifaceted public policy issue that includes important scientific, ethical, economic and political dimensions. Representing 181 countries, the OIE has a long-established role in setting global standards for animal health and broadened its activities to include animal welfare in 2001. The organization's vision is 'a world where the welfare of animals is respected, promoted and advanced, in ways that can complement the pursuit of animal health, human well-being, socio-economic development and environmental sustainability'. Using international experts, its animal welfare standards are based on scientific research, ethical considerations and practical experience. As well as developing standards, set out in its Terrestrial and Aquatic Health Codes, it implements them and supports training, awareness and communication. The OIE has adopted ten general principles for the welfare of farm animals (Fraser et al., 2013):

(1) Genetic selection should always take into account the health and welfare of animals.
(2) The physical environment, including the substrate (walking surface, resting surface etc.) should be suited to the species and breed so as to minimize risk of injury and transmission of diseases and parasites to animals.

(3) The physical environment should allow comfortable resting, safe and comfortable movement, including normal postural changes, and the opportunity to perform types of natural behaviour that animals are motivated to perform.

(4) Social grouping of animals should be managed to allow positive social behaviour and minimize injury, distress and chronic fear.

(5) Air quality, temperature and humidity in confined spaces should support good animal health and not be aversive to animals. Where extreme conditions occur, animals should not be prevented from using natural methods of thermoregulation.

(6) Animals should have access to sufficient feed and water, suited to the animal's age and needs, to maintain normal health and productivity and to prevent prolonged hunger, thirst, malnutrition or dehydration.

(7) Diseases and parasites should be prevented and controlled as much as possible through good management practices. Animals with serious health problems should be isolated and treated promptly or killed humanely if treatment is not feasible or recovery is unlikely.

(8) Where painful procedures cannot be avoided, the resulting pain should be managed to the extent that available methods allow.

(9) The handling of animals should foster a positive relationship between humans and animals and should not cause injury, panic, lasting fear or avoidable stress.

(10) Owners and handlers should have sufficient skill and knowledge to ensure that animals are treated in accordance with these principles.

The International Organization for Standardization (ISO) acts to align technical specifications with the principles of the OIE's Terrestrial Animal Health Code, ensuring the welfare of farm animals across the supply chain. The aim is to enable business operators to demonstrate their commitment to animal welfare by using a common approach, thereby overcoming or integrating the mass of private and public standards that business operators, especially in developing countries, may have to address (Tranchard, 2016).

There are also various independent groups, often charities, typically made up of a range of people, often experts, who review the evidence relating to animals and come up with recommendations. For example, in the US the Pew Commission's *Putting Meat on the Table: Industrial Farm Animal*

Production in America (2008) called for, among other things, the phasing out of intensive confinement systems that restrict animal behaviour, along with assistance for producers to make those changes. Similarly, among the recommendations in the UK's Food Ethics Council's *Farming Animals for Food: Towards a Moral Menu* (Mepham, 2001) were for changes in international law with the aim of improving farm animal welfare at a global level, including specifically exploring support to compensate farmers for the added costs of improved welfare measures. The Good Agricultural Practices developed by the United Nations' Food and Agricultural Organization address agriculture's social, environmental, economic and production goals, empowering farmers to farm. Their practices include the provision of adequate feed and water, avoiding non-therapeutic manipulations, minimizing live transport, handling animals with care and avoiding, for example, the use of electric goads, routinely maintaining animals in appropriate social groupings, and conforming to minimum space allowances and maximum stocking densities. Alongside animal welfare there are also expectations relating to soil, water, crop and fodder production, crop protection, animal production, animal health, harvest and on-farm processing and storage, energy and waste management, human health, welfare and safety, wildlife and the landscape.

Animals have been the subject of many modern philosophical treatises aiming to raise the status of animals and improve their welfare (Table 4.1).

There are a wide variety of animal advocacy groups, variously known as animal protection, welfare, rights, shelter, help or care organizations. While there were 500 in the US at the beginning of the 20th century, 100 years later the number had grown to 21,000, part of the increasing societal interest in subjects like the environment, human health and nutrition, and rural communities, as well as animals. Support for them has also increased; US animal societies now raise nearly 20 times the revenue per capita that they did a century ago (Shields et al., 2017). The diversity of the groups reflects in part their activities. For example, the US Animal Legal Defense Fund has been a driving force in the growth of animal law and the Humane Society of the United States active in getting citizen-led ballot initiatives to phase out the confinement of pigs in gestation crates and hens in battery cages. World Animal Protection has campaigned for grass-fed beef systems, advocating they can bring benefits for animals, farmers, the local economy and consumers, offering a commercially viable alternative to feedlot cattle production (Anonymous, undated, b). In opposing mega-dairies, the organization has also advocated for pasture-based dairy systems, producing a comprehensive economic appraisal of the differences between the intensive high-output cows and more 'robust'

Table 4.1 *A selection of some of the modern philosophical works addressing humans' relationships with animals.*

Work and author

Animal Liberation. Singer (1975)
Animals and Why They Matter. Midgley (1983)
Created from Animals. The Moral Implications of Darwinism. Rachels (1990)
Biology, Ethics, and Animals. Rodd (1990)
Animal Rights & Human Morality. Rollin (1992)
The Animals Issue. Moral Theory in Practice. Carruthers (1992)
Animals, Politics and Morality. Garner (1993)
The Political Animal. Biology, Ethics, and Politics. Clark (1999)
Animal Revolution. Changing Attitudes towards Speciesism. Ryder (2000)
Animal Rights. Current Debates and New Directions. Sunstein and Nussbaum (2004)
Animal Pragmatism. Rethinking Human-Nonhuman Relationships. McKenna and Light (2004)
Ethics of Animal Use. Sandøe and Christiansen (2008)
Veterinary and Animal Ethics. Proceedings of the First International Conference on Veterinary and Animal Ethics, September 2011. Wathes et al. (2013)
A Plea for the Animals. The Moral, Philosophical, and Evolutionary Imperative to Treat All Beings with Compassion. Ricard (2016)

grazing herds (Meaden et al., undated). Major food retailers are encouraged by Compassion in World Farming to demonstrate their commitment to improving farm animals, the charity recognizing those committed to sourcing product with higher welfare standards and avoiding lower standards. The *Business Benchmark on Farm Animal Welfare* assesses global food companies on their animal welfare policies such as close confinement and long-distance transport, how they govern and manage the supply chain, progress with changing, for example, to cage-free systems, and measures of animal well-being (Amos and Sullivan, 2016; Sullivan et al., 2017). Animal rights organizations, such as People for the Ethical Treatment of Animals or PETA, founded by Ingrid Newkirk and Alex Pacheco, focus on changing practices through, among other things, investigating and exposing cruelty, rescuing animals, seeking legislative changes, involving celebrities and protesting, sometimes provocatively. Other advocacy groups also expose inhumane practices, often through covert activities, and advocate for changes in societal attitudes. Recognizing the susceptibility of major brands and retailers to adverse publicity, some groups exert pressure for enhanced food supply chain practices and corporate social responsibility. Finally, some, such as the UK's Animal Liberation Front, have used more violent and illegal tactics (Henshaw, 1989).

Advocacy groups, be they industry or animal advocates, have at times attracted their own criticisms. For example, giving others the impression that they exclusively represent the truth, and that they merely have to 'educate' others, whereas when people do not have an opportunity to become involved and share their experiences, it can lead to resentment (Gore, 2007). Similarly, some are open to charges of 'sensationalising and leaving out inconvenient truths that fail to fit their agenda' (Norwood and Lusk, 2011). Finally, some corporate social responsibility-like initiatives determining how businesses should address animal welfare, are based on the expectation of significant influence on their supply chains. This expectation is troubling; what is that influence, what could, or should, it be and should responses be shared equitably across the supply chain? On the other hand, corporates have, among other things, incentivized farmers to provide high standards of animal welfare, supported research and development, and promoted higher welfare to increase consumer demand for higher animal welfare products.

Science has had an impact on animal welfare in two general ways. Firstly, animal production science has been instrumental in providing much of the knowledge and technology for the modern intensification of farming and its increases in productivity. For instance, science is now addressing the early life of broilers or meat chickens by targeting nutrition 'programming' digestion before the bird has hatched, *in ovo* feeding, to increase growth rates (Uni et al., 2005). In facilitating an increase in production, science has, indirectly or directly, helped create some of the problems of modern farming systems, for example production diseases, and is also attempting to mitigate them. While productivity and economic efficiency have been valued as a goal of science, the sustainability of the social-ecological system has tended to be ignored and scientists could ask: what public goods should agriculture serve? The second way science contributes is in understanding the impacts of farming on animals' lives (Table 4.2).

Animal welfare science grew out of the need to incorporate animal behaviour into assessments of animal well-being. In an appendix to the Brambell Report, Professor W.H. Thorpe (1902–1986) drew attention to two fundamental types of scientific evidence: physiological, the structure of the animal and the function of its parts; and ethological, its behaviour (Thorpe, 1965). Described by a colleague as a magpie science, animal welfare science collects bits of information from anywhere and encompasses disciplines such as behavioural ecology, cognitive science, endocrinology, genetics, health, immunology, neurophysiology, pathology and social science, often drawing on knowledge from a range of domestic and wild animals, and humans. There are, and have been, many internationally

Table 4.2 *A selection of some of the foremost works on farm animal welfare science and husbandry.*

Work and author
Animal Suffering. The Science of Animal Welfare. Dawkins (1980)
Livestock Behaviour. A Practical Guide. Kilgour and Dalton (1984)
Stress and Animal Welfare. Broom and Johnson (1993)
Animal Welfare. A Cool Eye towards Eden. Webster (1994)
Farm Animal Welfare. Rollin (1995)
Animal Welfare. Spedding (2000)
Physiology and Behaviour of Animal Suffering. Gregory (2004)
Animal Welfare. Appleby and Hughes (2005)
Animal Welfare: Limping towards Eden. Webster (2005a)
Understanding Animal Welfare. The Science in its Cultural Context. Fraser (2008)
The Welfare of Sheep. Dwyer (2008)
The Sciences of Animal Welfare. Mellor et al. (2009)
Animals Make Us Human. Creating the Best Life for Animals. Grandin and Johnson (2010)
Improving Animal Welfare: A Practical Approach. Grandin (2015)
Domestic Animal Behaviour and Welfare. Broom and Fraser (2015)
Animal Welfare in Extensive Production Systems. Villalba (2016)

renowned researchers and teachers (Broom, 2011; Newberry and Sandilands, 2016); among them were Hans Seyle, David Wood-Gush and Ron Kilgour.

Selye (1907–1982) was born in Vienna and lived in Hungary, Czechoslovakia, the US and Canada. A medical doctor with a PhD in chemistry, it was his research as an endocrinologist that led him to demonstrate scientifically the response of the body to biological stress, the ability to survive life-threatening situations. Initially injecting extracts of organs in a search for hormones, he recognized consistent pathophysiological responses. Selye noticed that when exposed to different insults such as exposure to cold, surgically induced injuries, excessive exercise or sub-lethal doses of drugs, rats responded in the same way: lymph node and thymus atrophy, gastric ulcers and enlarged adrenal glands (Table 4.3).

Initial decreases in organ size and other changes were followed by enlargement of the adrenal gland and reduction in growth and production, the animals eventually succumbing if unable to adapt. As the response seemed 'to represent a generalized effort of the organism to adapt itself' to the insult, Selye termed the response a general adaptation syndrome, later renaming it stress. The three phases of the response were: alarm, resistance

Table 4.3 *The General Adaptation Syndrome described by Selye (1936) based on common symptoms observed in rats after exposure to different noxious agents.*

Stage	Symptoms
Alarm, 6–48 h after injury	Decrease in size of the thymus, spleen, lymph glands and liver; disappearance of fat tissue; oedema formation, especially in the thymus and loose retroperitoneal connective tissue; accumulation of pleural and peritoneal transudate; loss of muscular tone; fall of body temperature; formation of acute erosions in the digestive tract, particularly in the stomach, small intestine and appendix; loss of cortical lipoids and chromaffin substance from the adrenals; and sometimes hyperemia of the skin exophthalmos, increased lachrymation and salivation. In particularly severe cases, focal necrosis of the liver and dense clouding of the crystalline lens are observed.
Resistance, beginning 48 h after injury	Adrenals are greatly enlarged but regain their lipoid granules, while the medullary chromaffin cells show vacuolization; the oedema begins to disappear; numerous basophiles appear in the pituitary, the thyroid shows a tendency towards hyperplasia; general body growth ceases and the gonads become atrophic; in lactating animals, milk secretion stops.
Exhaustion, 1–3 months	If the insult is severe and prolonged, animals lose their ability to adapt (as in the Resistance stage) and display similar symptoms to the Alarm stage and succumb.

(habituation and inurement), and, if unable to adapt, as the ability to resist was finite, exhaustion.

Selye noted that anything that causes stress, whether physical, toxic, chemical, psychological or the result of a pathogen, endangers an organism's life unless it can respond adequately. Thus, adaptability and resistance to stress are fundamental to good welfare. The animal response was coordinated through the nervous (e.g. release of the hormone adrenaline) and endocrine systems (e.g. release of the hormone cortisol) and involved almost every other organ, especially the cardiovascular, pulmonary and renal systems, and even cognitive function. Selye demonstrated the crucial role of the pituitary–adrenal cortex axis in the stress response, the production of cortisol helping regulate metabolism during the alarm phase where the body adapts quickly to be able to flee or confront a threat. He also named glucocorticoids and mineralocorticoids that regulate carbohydrate and mineral/electrolyte metabolism and exert anti- and pro-inflammatory effects.

An animal's susceptibility was dependent on its genetics, experience, exposure to disease, and diet. Seyle differentiated between the damage caused by the agent, and the defence against it. Up until that time, medicine addressed almost exclusively the agent, trying to neutralize the insult. The general adaptation syndrome suggested that, as well as advising rest and a wholesome diet, medicine could assist adaptation to injury by supplementing suboptimal natural defence mechanisms. Seyle later introduced the terms distress and eustress to distinguish between the stress responses to negative or unpleasant, and positive or pleasant feelings or, as one commentator described it, the brain recognizes the difference between arguments with your spouse and kissing one's girlfriend or boyfriend. Selye defined stress as the non-specific response to any demand placed upon an organism, reserving the term distress for the unpleasant or damaging impacts of stress.

Arguably one of the most creative scientists of the 20th century, Selye was a world authority in endocrinology, steroid chemistry, experimental surgery and pathology reflected in 1500 articles and 32 books, including his best-seller, *The Stress of Life* (Perdrizet, 1997; Szabo et al., 2012). He was nominated several times for a Nobel Prize. According to one account, Selye's legacy was also apparent in his belief, 'theories don't have to be correct – only facts do' and in the motto, 'neither the prestige of your subjects and the power of your instruments, nor the extent of your planning can substitute for the originality of your approach and the keenness of your observation.' He believed theories encouraged the discovery of new facts, and, in turn, led to better theories. Subsequent research has revealed that the stress complex is much more variable than Selye theorized, and it can be specific to the insult. Furthermore, single adrenal indices can be poorly correlated with the adverse effects of insults, something many tend to forget in the almost singular belief in the 'stress hormone' cortisol. Selye may also have had an involvement with the tobacco industry, receiving extensive funding and it is claimed 'his research on stress and health was used in litigation to defend the industry's interests and argue against a causal role for smoking in coronary heart disease and cancer' (Petticrew and Lee, 2011).

One of the first to study the impacts of factory-farming on animal welfare was the South African-born geneticist and ethologist David Wood-Gush (1922–1992). Wood-Gush advocated that to understand animals, and their welfare, you had to understand their behaviour, believing that animal welfare standards were often based on little real knowledge of behaviour but were arbitrary and with a 'good sprinkling of anthropomorphic sentiment'. Consequently, he championed the design of housing systems catering for the animals' needs based on studies of how they use a more natural environment. Wood-Gush held that humane treatment was determined by balancing the impacts of interventions on behaviour against the benefits for humans, frequently economic. He also held that animals should be treated as individuals not commodities. His position was to present the pros and cons of things like different housing systems from which politicians and their electorates could choose the level of welfare they could adopt while protecting their farmers.

Wood-Gush also advocated for more studies of the impacts on animals of chronic as opposed to acute stress, for example, the performance of stereotypies, short sequences of behaviour or a single behaviour pattern repeated over and over, for no obvious reason. For example, bar-biting and head-weaving, common in impoverished environments, indicate the animal is severely frustrated being unable to manage its circumstances. In more enriched environments, such invested novel and repetitive behaviours cease or are reduced, although not always (Mason and Latham, 2004). Animals in richer environments may be better able to cope with some novel conditions, but less reactive to some environmental changes. For instance, frustration results in aggression in hens, and pigs, who have a high level of curiosity, scamper and play and explore novel objects, underlying the inadequacy of barren industrial-like housing systems (e.g. Duncan and Wood-Gush, 1971; Wood-Gush and Beilharz, 1983; Wood-Gush and Vestergaard, 1989, 1991).

However, it was Wood-Gush's method of studying the behaviour of animals in a variety of environments, both ecological and social, which enabled appreciation of the full repertoire of behaviour that provided an insight into motivation and the control of behaviour (Stolba and Wood-Gush, 1984). Concerned with how chickens responded to very intensive husbandry, he had birds released on an island off the west coast of Scotland to study their social organization in the absence of the constraints of captivity. Unfortunately, the birds succumbed to the island's mustelid predators. Undeterred, the behaviour of pigs was observed in a semi-natural wood and grassland enclosure, the Edinburgh Pig Park, containing a marsh, a stream and bushes, at the Edinburgh School of Agriculture (Figure 4.2).

The sows built large nests, selected sites with shelter and views, were highly sociable and the young ones played a lot. The pigs' behaviours were compared with those in less complex environments (paddocks, yards and a range of pens). Animal behaviour was guided by specific features consistently present when behaviours were performed, for example, nesting in the farrowing sow, dedicated defecating areas and close relationships between adults and between juveniles. With Alex Stolba, a novel enriched housing environment, the Family Pen System, was designed, reproducing the features guiding the pigs' behaviour, some of which were only observed in the

Figure 4.2 The Edinburgh Pig Park, a semi-natural enclosure in which the natural behaviour of domestic pigs was studied (*Photo: Alex Stolba and David Wood-Gush, courtesy of Alistair Lawrence*).

semi-natural enclosure. The features of each compartment of the enriched communal system included:

- roofed and open components;
- a feeding area separated from a resting area;
- a farrowing nest positioned against the inner pen walls;
- space large enough for nesting;
- an outdoor corridor for defecating away from the nest;
- a rooting area;
- a rack for gathering nesting materials;
- a rubbing and marking post;
- a partitioned feed trough, facilitating undisturbed individual feeding;
- partitions allowing pigs to hide from others; and
- access to the other compartments, supporting a species-typical social structure.

The system was based on a small and stable social grouping in which young pigs were often naturally weaned and remained until sold, removing the need for mixing groups and the need for specialist housing common in modern piggeries. Each unit housed four sows, each with her own compartment but able to move freely and interact with the other sows and their piglets, as well as with the varied resources the system provided.

What the programme aimed to achieve was to understand the animals' natural behaviours and then design a system in a more confined space that enabled the animals to still undertake those behaviours. Good welfare was achieved when no specific motivation to behave remained unrelieved, the environment satisfying the different aspects of a behaviour, for example, the physiological (nutritive) and behavioural (foraging) aspects of feeding. Similarly, maintaining the pigs' basic social structure, while living in a confined environment, lessened the problems usually associated with confinement by giving the animals the opportunity to express their behavioural repertoires. Failure to acknowledge the different appetitive (e.g. feeding) and exploratory phases (e.g. foraging for feed) can result in pigs remaining frustrated despite being, for example, well-fed. Frustrated animals direct their behaviour to their penmates, often resulting in non-nutritive sucking and injurious behaviours (e.g. feather pecking, tail-biting and cannibalism). The animal should feel that it has some control over its environment, expressed in meaningful appetitive behaviour (e.g. foraging), or be able to predict parts of it (pigs fed at irregular times, thus unable to predict or control feeding, show higher levels of aggression). Although there have been problems with the prototype family pen system, for example, piglet deaths and increased need for resources, subsequent versions have sought to overcome them resulting in practicable commercial systems (Wechsler, 1996).

Another pioneer of animal welfare science was Ron Kilgour (1932–1988), considered the father of farm animal behaviour and its role in animal welfare in New Zealand (Figure 4.3). Trained in psychology, Kilgour's popularity with farmers stemmed from his focus on addressing practical farm animal issues, an admiration perhaps envied by other more traditionally dairy farm-trained scientists and their leaders of the day. While presenting at international conferences and publishing in the science literature (e.g. Kilgour 1975, 1978), it was the information he produced for farmers for which he was most well known (e.g. Kilgour, 1972, 1976; Figure 4.3). He also had a tremendous liaison with university students, pulling people out of the audience and saying 'you're the dog, and you're the owner' and inviting the latter to write a 'domestic contract' for the former.

Kilgour examined aspects as diverse as the behaviour of ewes at lambing, cow's food preferences, humane handling of stock prior to slaughter, the domestication of deer, and the role of human–animal bonds. 'The regular routine, so essential for the cow, means a very ordered and perhaps boring life for the milker, so the temperament of the milker becomes as important as that of the cow – a sobering thought.' Humane care was understood as maintaining husbandry procedures in keeping with an animal's traits; farms needed to be fitted to animals, as well as animals fitted to farms. Knowledge of species-specific behaviour was seen as essential to

Figure 4.3 Ron Kilgour, an animal psychologist and behaviourist who believed that you should 'never do for an animal what it can learn to do for itself' *(Courtesy Glen Kilgour and the Ron Kilgour Memorial Trust)*.

good animal husbandry. For example, many farmed species are altricial at birth, mother and young having to quickly bond if the young are to survive. He identified the importance of the physical birth site, to which ewes were attached, and its smells, something shepherds were well attuned with. Farm animals are also social animals, living in herds and flocks where violations of social space, especially in intensive environments, can disturb them. 'Milking, to be profitable and satisfying, should be a cow-centred activity. Anything which interferes with the wellbeing or upsets the routine of the cow lessens the chances of obtaining maximum yield.' Mismothering can be a significant problem when the development of the social bond between mother and young is disturbed in the first few hours after birth, many animals preferring to give birth in isolation, returning to the flock or herd after several days. Knowledge of the impacts of husbandry on animal behaviour was also important; the expression of oestrous behaviour in housed dairy cattle is most common at night in the absence of feeding and milking schedules. Kilgour noted factors as diverse as bureaucratic regulations, poorly designed yards, lack of space, disease and transport that could be distressful to animals and questioned how other practices, such as strip-grazing, pubertal mating, abrupt weaning and giving animals the opportunity to learn of their environment from their dams (hefting), might be interpreted as humane.

Kilgour was a believer in viewing the world from the animal's perspective. He had people stand amongst a group of cows in the paddock, yard and milking shed, with eyes at cow level to see and feel what the cows saw and felt. In one study observing the impacts of the transport of 3–4-month-old calves by road and air (Kilgour and Mullord, 1973), observers travelled the 1600 kilometres over four days with the animals. After 13 hours of continuous travel, 60% of the calves were either lying down or grinding their teeth, with signs of fatigue: heads hanging low, legs straddled and eyes half closed. The importance of adequate head room during travel was also noted, those without it tending to lie down and risking being trampled. When the calves were released onto pasture midway through the journey, their prime need was for exercise, with grazing next and water only a third consideration. They ran, grabbed a bite of pasture, and ran again for almost 1.5 hours after having been penned in for 25 hours. The daily activity patterns displayed by the calves raised the possibility of transport between grazing cycles, easing the necessity for, and pressure of, travel during darkness on both animals and drivers. Kilgour considered that standards of humane care should be set by consensus and achieved by avoiding injury, handling animals in the least stressful manner and meeting their species-specific needs. He also considered that it is important to remember that animal welfare is a total society responsibility, modern concerns about intensively farmed animal welfare rightly part of a re-examination of the terms of the domestic contract or the trade-off implicit in domestication.

Science is one of our more powerful means of understanding and predicting the world, including animals and people. However, and like anything else, there is 'good' and 'bad' science, and it can be used to 'sell' or 'defend' any position. As experts have an influential position in society, the scientific community may need to devise a way by which it can swiftly and publicly provide context for, or rebut, claims that have questionable scientific foundation. Despite its value, and its somewhat exalted position in animal welfare, science cannot be used as a substitution for values, the real determinant of responsibilities to animals. As Winston Churchill held: 'scientists should be on tap, but not on top'. It is important that the uses and limitations of any discipline or function providing insights into animal treatment are acknowledged. And, the choices people have to make are also unfortunately unacknowledged, for example, that effort invested in one intervention inevitably means less effort going on other things. People are rarely swayed by scientific facts, but by values, stories and context. Science tells us what is, whereas ethics, a generalized set of 'rules' for human behaviour, assists with what we should do. The limits of relying on science, or any evidence, are recognized; while they can help with understanding animal welfare, they are not a substitute for solving ethical, social and economic problems that make animal welfare a complex and much-contested subject. For example, the development of enriched cages with nesting areas, scratching strips and a greater area allowing more movement for laying hens, may not adequately address the concerns associated with the conventional battery cages they sought to replace, resulting in enriched cages, possibly because they are still cages, also being contested (Doyon et al., 2016; Weary et al., 2016). Science must address the concerns people have, because failure to do so questions the legitimacy of the science and the profession. Many of the issues confronting agriculture are social and ethical and yet emphasis is placed on researching and teaching the reductionist or mechanistic science, perhaps because it is easy, and it allows scientists to publish, to maintain credibility and to advance their careers. However, what does something like the number of hormone receptors in the brain really mean? The real problems for agriculture are much harder, including ecology, the environment, business planning and family relationships. These are more difficult to teach but may well make the biggest impact on farm welfare and the future.

Many animal industries and food retailers have developed expectations of their suppliers including farmers. Such expectations are frequently articulated in policies and assurance schemes, mostly merely reflecting legislation, and often audited. Originating in demands on retailers to ensure food safety, consumer concerns for food quality and animal welfare, or in farmers wanting to assure compliance with standards or differentiate their systems, the food supply chain is awash with product quality labels, claims and assurance schemes. Crises, like mad-cow disease, horse-meat sold as beef, and animal welfare exposés and vocal interest groups, prompted major food retailers and fast-food outlets to take a greater interest in food production. Farm assurance covers things like product quality, traceability, hygiene and food safety, animal welfare, crop, field, water and soil management, and the competence of people involved. Arguably, there is also a desire, on the part of major food retailers, to standardize food, increase control over the food supply and increase their profit margins, rather than or as well as establishing farm assurance.

The breadth and variety of such schemes are staggering. For instance, in the US in 2011, there were 56 eco-labels for beef ranging from organic to animal welfare-approved, humanely raised and handled, free range, grass fed, to no additives and hormone-free. There were also 48 such labels for dairy, 46 for eggs, 50 for lamb, 49 for pork and 54 for poultry (Niles, 2013). Some animal welfare labels' claims were based on formally defined standards with compliance by third-party audit (e.g. certified humane, organic and grass-fed). Others had no legal definition, were based on vague or weak standards and had unverified compliance (e.g. free range, natural raised, sustainably farmed). And some were meaningless or misleading (e.g. natural, no added hormones, vegetarian-fed), being more for marketing purposes (Animal Welfare Institute, undated). Some schemes follow legislative minimums, others reflect higher standards. One comprehensive example is the EU's Welfare Quality scheme assessing feeding, housing, health and behaviour (Table 4.4).

Supply chains and retailers are competitive and assurance programmes have a number of roles ranging from assuring consumers, to distinguishing retailers from their competitors, and from shifting the responsibility for product safety and quality to others in the supply chain, to reconnecting consumers with farmers. While they may provide customers with choices and producers with premiums, their sheer number and complexity may confuse consumers and impose additional costs on producers, especially if they are not being recompensed for higher standards (Early, 1998; Food Ethics Council, 2008; Thompson et al., 2007), or they may not lead to demonstrable improvements in animals' lives (Webster, 2009). At present the impacts on animal welfare of such schemes are varied or inconclusive. Though membership is associated with a greater likelihood of complying with animal welfare rules and standards, perhaps the better farmers are more likely to join them (see Hubbard, 2012).

The goal of assessing or auditing farm animal welfare is, too, a complex subject. What to measure, how often, and on what proportion of the herd or flock? What is acceptable repeatability between observers, and what is acceptable in different systems (e.g. flight distances in extensively and intensively farmed animals)? How are different measures weighted and can deficiencies in one measure be compensated for by adequacies in

Table 4.4 The four principles and 12 criteria, along with examples of measures, of the Welfare Quality on-farm animal welfare assessment system (Blokhuis et al., 2013) and measures for the welfare assessment of growing pigs (Temple et al., 2011).

Principle and related criteria	Examples of measures	Measures used in a study on pig farms
Good feeding		
Absence of prolonged hunger	Body condition score, feed management, placement and maintenance of resources	Body condition score
Absence of prolonged thirst	Number and type of water bowls and drinkers, alarms	Water supply
Good housing		
Thermal comfort	Shivering or panting, temperature and humidity, air flow	Shivering, panting, huddling
Comfort around resting	Cleanliness of animals, litter quality, time needed to lie down	Bursitis, absence of manure on the body
Ease of movement	Access to outdoors and pasture, slipperiness of floor, stocking density	Space allowance
Good health		
No painful management procedures	Prevalence (e.g. beak trimming, tail docking), use of pain relief	Castration, tail docking
No disease	Nasal discharge, milk somatic cell count, scouring, apathy, records, culling rate	Coughing, sneezing, laboured breathing, twisted snouts, rectal prolapse, scouring, skin inflammation or discolouration, ruptures, and hernias
No injuries	Incidence of lameness, wounds, dermatitis, protection from predators	Lameness, wounds on body, tail biting
Appropriate behaviour		
Expressing social behaviour	Agonistic and aggressive behaviours, enrichment measures, grooming	Social behaviour
Expressing other behaviour	Playing, exploratory behaviour, use of resources (e.g. nests)	Exploratory behaviour
Good human–animal relationship	Flight or avoidance distances, reaction to humans, stockman inspections	Panic response to humans
Positive emotional state	Assessment of body language (qualitative behaviour assessment)	Qualitative behaviour assessment

others? Should they use environment and management (e.g. space allowance) and/or animal-based (e.g. animal behaviour) measures? How practical are they and what time is required to undertake them? These and other questions determine the form the assessments take (Johnsen et al., 2001). There are also different understandings and interpretations; an animal welfare inspector or legislative perspective may differ from the context provided by farming animals. Having ownership and responsibility for animals is crucial for being motivated to improve the farming system and thus animal welfare. Having the confidence, time and resources to do what they are good at is undermined by inspections not being cognizant of the context, the features of the farming system and why animals in a system are treated as they are (Anneberg et al., 2012). Perhaps there is an opportunity for farmers to have a greater role in setting standards or at least leading public debates rather than having to defend their practices when 'told' how to farm. Despite these concerns, auditing has been used to improve and maintain high standards of, for example, handling and stunning at livestock slaughter plants in the US (Grandin, 2013). Measures have included the correct placement of electrodes for stunning and the proportion of animals rendered insensible with one stun, and the minimal use of electric prodders. Other concerns include the credibility of standards, their potential for use as discriminatory trade barriers, their multiplicity, compliance costs and lack of consumer (or societal) input (More et al., 2017). Do consumers really want information or just to be able to put their trust in the system, to trust that animals have been given a good life and a quick death?

Reflection on the past, present and future of farm assurance schemes unsurprisingly painted a positive picture of farm assurance schemes, making animal welfare standards explicit and a foundational part of what they do, providing confidence that standards are in place, and helping to continually drive higher standards of welfare. While there is potential for such schemes to become more engaged and influential, perhaps even seeing themselves as co-custodians of farm animal welfare, there was also a need to address their costs and to balance the requirements of consumers, farmers and retailers. Furthermore, farm animal welfare should be considered as part of a fair, healthy, humane and environmentally sustainable food and farming system. For instance, the health and well-being of the people handling and caring for animals also needs to be considered, a vitally important issue all too often ignored (Food Ethics Council and Pickett, 2014). The risk is that unless considered from a whole of system perspective, these schemes, which may or may not protect or enhance welfare, and may or may not reassure the consumer, will impose additional costs without commensurate rewards.

Ideally, assurance schemes should provide assurance to customers, but they are also perceived as a 'necessary evil' by farmers, and a 'gate-keeping' device for food safety and quality by retailers (Hubbard, 2012). Health planning, a requirement of some farm assurance programmes, assists with things like mastitis and lameness but, again, telling may not be as good as people coming to it in their own time. It also remains to be seen whether legislative mechanisms will be necessary to manage the authenticity of the claims made by private or market-based systems. Market-based solutions only appeal to consumers of products, not those who do not consume but who inevitably have concerns about animal welfare, some of whom may not consume products because of concerns for animal welfare. Though non-consumers, for example, vegetarian animal advocates, may drive them. Consequently, legislation is required to give voice to those people.

The final part of the food supply chain is the consumer, arguably the 'greatest practical obstacle to the improvement of farm animal welfare', being reluctant 'to convert their oft-expressed desire for high standards of animal welfare into a demand for high-welfare goods when they enter the food store' (Webster, 2005b). In this view, consumers, who drink milk, eat meat and eggs, and wear the woollen and leather products of animal farming, also have a responsibility for animal welfare since, as the philosopher Kant claimed, 'he who justifies the ends justifies the means also'. The choices of what to eat and to wear, and how much to pay, ultimately, along with the decisions others in the supply chain make, help determine what happens to animals and farmers.

Affluence has brought both public scrutiny of farming practices and a belief in the sovereignty of the consumer, with any demand for quality met as a king may demand loyalty of his subjects. Kings, like consumers, are varied in their interests, character and wisdom. For example, Solomon, an ancient King of Israel (961–922 BC), was the wisest of the wise. The Judgement of Solomon tells of him proving that the mother of a disputed baby was the one who offered to give up the child rather than have it cut in half. In contrast, Herod the Great, who lived 900 years later as King of Judea, was a tyrant held in loathing for his killing of all the male babies in Bethlehem to kill the infant Jesus. If the consumer is to be king, which king is to be representative: Solomon or Herod?

So it is with consumers who differ in their attitudes, opinions, values and sentiments, in time and space, in the belief that their choices matter or needs must be heard, or the understanding that animal welfare is or is not their responsibility. Consumers are far from a homogeneous or generic group and, despite being concerned for animals' welfare, some have a poor understanding of the realities of modern farming (Cornish et al., 2016). They can be confused and misinformed of issues, may not want to know about how animals are kept or food produced, find it difficult to become informed, especially when shopping, and their beliefs are often based on personal experience with animals, the media, animal advocacy groups, magazines, radio, newspapers, friends and family, internet, and mass media images of negative issues. The public and media scrutiny and discussion undoubtedly contribute to attitudes and opinions and animal advocacy organizations are an important source of information. Interestingly, consumers seem to recognize that farmers alone are not responsible for compromised farm animal welfare, but that profit-driven industry and demand for cheap food without consideration are also implicated (Spooner et al., 2014).

While supporting high standards of animal welfare (e.g. the provision of food and water and prevention and treatment of injury and disease), consumers vary in how they value other aspects. To some, it was also important to raise animals in a way that would keep food prices low, to others having animals able to behave normally and exercise outdoors was important (Prickett et al., 2010). Animal welfare is of secondary importance compared with the role of meat as food, its price, and whether it was produced locally (Autio et al., 2018). Finally, and as indicated in a survey of US households in 2007, while animal welfare is important, it is not all-important; human poverty, the health care system and food safety were each five times more important than animal welfare. Even the financial well-being of US farmers was considered twice as important as the well-being of farm animals (Lusk and Norwood, 2008).

While there is a belief that improving animal welfare is a role that consumers should undertake, expectations of active, ethically conscious consumers are fraught with difficulties (Jacobsen and Dulsrud, 2007). These include consumers in different countries seeing their responsibilities and powers differently, lack of reliable information on which to make decisions, and the moral complexities of everyday life that restrict such a role. Some may prefer not to think about it, a psychological mechanism for dealing with harming animals (Chapter 1), otherwise knowing an animal died 'makes me cry when I cook a meal for the family'. Others may believe good animal welfare means better taste and protects or improves human health. Or that it should not cost more and so not put farmers out of business. Furthermore, consuming animal products supports farmers and benefits the economy. Many consumers are also aware of the structural factors, mostly economic, which force producers to farm in the way that they do. They also recognize that they, the consumers, are part of the problem as by purchasing cheap products they sustain the market for conventional systems: '... when I look at meat in the supermarket, I don't read about the way it is produced, I don't really care about it ... I hate to admit it, but that is the consequence I actually don't care!' (Lassen et al., 2006b). Furthermore, many people choose not to think about the way their food is produced, the way the animals are farmed, preferring to remain, wilfully, ignorant because they trust farmers and/or have more important things to deal with, or perhaps to avoid any guilt associated with the death and use of animals (Bell et al., 2017). It may partly explain the difference between willingness to pay and actually paying for or supporting animal welfare-friendly products. Furthermore, if consumers do trust farmers, and are relatively uninformed of farm practices and animal well-being, supporting farmers to care for animals may be the best strategy. However, it is a complex issue and it is not known 'whether greater wilful ignorance benefits or harms livestock'. Accepting consumers 'forget' about animals when paying for animal products, a producer's view in the British Society of Animal Production's *Farm Animal Welfare – Who Writes The Rules?* was that it would always be a dilemma until there was a demonstrable relationship between animal welfare and product quality that people were willing to pay for. It was this reluctance to pay that will continue to restrict progress in animal welfare (Lowman, 1999). Perhaps we have still to explore what role could, or should, consumers be expected to undertake? Do consumers really want to know that the meat came from an animal? Or trust the system that it was cared for? And how do consumers see themselves as needing to be responsible when, arguably, other parts of the supply chain should play a central role? Is it necessarily wrong to eat food from animals that have not had a good life? The consumer must be persuaded that the value is real, justified and worth it, and the farmer is rewarded, otherwise it is just another cost of production.

Enhancing animal welfare, especially in intensive systems, incurs higher production costs (Chapter 3). Is there a market for products claiming to have higher standards? Are consumers willing to pay for higher animal welfare products? These questions have been examined by asking how much people are willing to pay for animal welfare. Overall, consumers are willing to pay (Clark et al., 2017; Lagerkvist and Hess, 2011; Norwood and Lusk, 2011) a small premium for animal welfare, increasingly so with income and education, less with age, and higher for females. While increased willingness to pay applied to all animals, it was lowest for pigs and highest for beef cattle. However, there are limitations with the methodology, most notably consumers expressing preferences as citizens when asked about the value of higher animal welfare but these preferences not being translated into purchasing decisions ('citizens desire but consumers dictate'). Animal welfare is rarely a priority when shopping compared with product presentation, price and quality, for example. And there is much variation in willingness, suggesting opportunities for niche markets and that different segments of the market with different preferences need to be catered for. Willingness to pay studies also depend on whom you see as being responsible for animal welfare. If it is the government or the farmer, then willingness to pay for enhanced animal welfare is low. People also see themselves as supporting animal welfare in their role as citizens, for example supporting legislators, and as not needing to do so again through their purchasing behaviours. In addition, as animal welfare comes under greater public and media scrutiny, consumers may be sensitized and then fatigued by, or resistant to, the increasing exposure and expectations. As willingness to pay is significantly dependent on knowledge about farm animal conditions, and how processed the product is, perhaps the most significant concern is that consumers do not have the necessary knowledge to make an informed decision. With few people now having direct contact with farm animals, public and social media, stories and celebrities, help portray human relationships with farm animals. Societal realization of the inhumane nature of some uses and treatment of animals, is often driven or reinforced by media exposés and disasters and the trade bans, public backlash and political embarrassment that accompany them. The provision of information rarely changes people's views. In fact, providing information on contentious issues may further polarize people; beliefs help determine actions, not more facts. Information is interpreted with emotions, instincts, experiences and life circumstances as well as knowledge. It is important to remember that the more we use media to understand the world, the more we will condemn ourselves to stereotypes, requiring the media to take a more sophisticated role in informing (as well as entertaining) us. For instance, rarely does a single scientific study stand as definitive proof and it needs to be portrayed within the context of the greater whole.

It is difficult not to conclude that willingness to pay, even consume, is not necessarily a good indicator of the acceptability of a production system though with sufficient labelling and information, traceability and verification and monitoring, market-based solutions may improve farm animal welfare. They will certainly change it. Much of consumption is mundane, everyday life can demand consideration of many things and geographically distant farm animals, and lack of reliable information, limit seriously questioning the moral burden consumers might be expected to carry. It has been suggested that society should consider the rightful place of private and public responsibilities towards animal welfare (Jacobsen and Dulsrud, 2007; Lundmark et al., 2018).

Finally, in return for year-round produce, uniformity and consistency, convenience, choice cuts and lower prices, we might ask, what has the consumer brought to the relationship between humans and animals?

The depth of their concern may surprise. However, given the plethora of schemes and motivations for establishing them, it is tempting to suspect that most consumers would find it difficult to choose, and consequently default, basing most of their decisions on habit, price or media attention. To conclude, perhaps it is another King, Henry VIII, who might better represent the consumer. Compared to the two earlier Kings, Henry could be considered as basically sound, but a bit rough around the edges. His good looks, hearty personality, fondness for sport and the hunt and military prowess endeared him to his subjects, at least initially. However, it is for his efforts in consummation that he is best remembered. Henry's search for a 'quality' heir to the throne resulted in six marriages, two wives executed and the creation of the Church of England. His demands for 'quality' have had some pretty far-reaching consequences. The modern consumer's demand for quality food may have equally far-reaching consequences. Like other parts of animal welfare, the consumer's place is complex and wicked.

The diversity of influences in this chapter, both historical and contemporary, bring their own insights and problems, presenting a challenge for society in determining what is acceptable use of animals. While the primary influence on the treatment of animals is through the actions of farmers, the husbandry ethic, there are two general ways for other individuals and their communities to collectively influence the welfare of animals. One is through the state, that is, legislation (institutional or collective ethics); the other through the marketplace, that is, individual or consumer ethics reflected in the rise of private standards, animal welfare labelling schemes and expectations for corporate social responsibility. However, and as noted in the current chapter, there are a myriad of ways in which both individuals and institutions can attempt to influence and constrain animal use, through the market or through legislation, reflecting and shaping our views of the morality of farming, both having advantages and limitations (Table 4.5).

Nowadays, market forces are becoming more influential, shifting the focus from citizens or their representatives, to retailers and consumers, changing animal welfare from a public to a private good. Along with scientists and philosophers, regulatory bodies, the media, educational groups, global governance organizations and private sector organizations including transnational corporations also help shape the morality of agriculture. Determining the morality of aspects of agriculture can only really be achieved by making provision for considering all those interested parties' views, acknowledging their insights, preferences and prejudices.

There are two themes apparent in these influences: standards can increase the cost of animal production; and while there is an expectation the consumer will pay, the number willing to do so, and the value

Table 4.5 *Some of the advantages and disadvantages of legislative and market-based approaches to animal welfare, suggesting both public and private standards have a role in maintaining and enhancing animal welfare.*

Legislation

Advantages	Disadvantages
Mechanisms such as codes of welfare have, in addition to their regulatory function, other roles: they articulate the aspirations of society, raise awareness of the issues by drawing attention to them, give society an idea of what is expected, and are self-promoting in defining a farming system and excluding others.	Risks of increasing interventions increasing bureaucracy and costs, and the state accumulating responsibilities. Similarly, risks inspiring respect for the law governing animal welfare rather than necessarily promoting a proper relationship with animals (Forsberg, 2011).
A single standard and source of authority makes it clear to farmers how they are expected to care for their animals.	Potentially open to the interests of powerful lobby groups at the expense of individuals and minority interests.
Acts to prevent practices that are, financially or economically, rational, i.e. correcting market failures.	Tendency for visible issues, those attracting public attention, to be addressed.

Market or private

Advantages	Disadvantages
Enables retailers to set different standards, including those above legislative requirements.	Consumers may want to avoid responsibility by 'delegating' it, see it as the responsibility of others or not care.
Swift, compared with legislation, retailers can make changes very rapidly, though requires transparency: who says, who pays?	Different standards make it difficult for farmers, e.g. multiple inspections can be required, increasing the burden on farmers.
Allows consumers to have a say in animal welfare, though not all concerned people contribute to the marketplace.	Risk that (especially retailer-driven) systems focus on consumer expectations and perceptions rather than animals' needs, and appealing and easy-to-communicate aspects rather than animal welfare improvements.

they attach, are limited. In other words, animal welfare is not just about farmers doing the right things, but everyone, businesses and consumers: 'if compassion towards animals is going to prevail, it must triumph in the market place' (Pacelle, 2016). The paradox is that the consumer cannot make that decision. And in leaving it to others 'the cost of good governance is soaring' (Singh, 2005). It is complex; the equitable distribution of animal welfare costs and benefits is not just a role of legislation, it is also 'about changing minds and moving markets' (Sandøe and Jensen, 2013). The various groups and their approaches to animal use can be simplified or generalized as based in the benefits of individuals using animals as a resource (i.e. economics) and the way we go about ensuring those resources are distributed fairly (i.e. ethics).

Ethics and economics

Values are individual and shared beliefs that motivate us and guide our sense of right and wrong, and our outlooks, aesthetics, choices and whether we view the treatment of animals as acceptable or unacceptable. They derive from many sources: our parents and family life, friends, religion, our trades and professions, and literature, and the media, in other words the society we live in and its history. Our common morality holds that we not harm, do good, are fair and just, and respect individuals' autonomy. When values are not shared, different views can be contentious. For example, the increased emphasis on values of efficiency and productivity associated with intensive farming systems contrasts with the more traditional values of animal husbandry and farming as a way of life. It is also apparent in the tension between the values of empathy, intimacy or affection for animals while at the same time relying and depending on them for labour, food, recreation and so forth. Many social factors further influence the tensions and inconsistencies, among them:

- the personal mores, skills and abilities, and needs of those people most familiar with, and responsible for, the welfare of animals;
- the varying importance of contact with animals in our lives and their removal, except for pets, from many people's lives associated with specialization and urbanization;
- the 'loss' of ancient or traditional ways of dealing with the 'psychological' tension between using and having empathy with animals;
- the legitimacy of different perspectives (pluralism);
- an understanding that ethics is the systematic evaluation of moral issues in the public sphere; and
- increasing numbers of people not having an awareness, or good understanding, of farm animal production, or the needs of animals on which they base their assessment of the ethics of farming practices.

Human values, then, are complex, variable and unpredictable; we do not always act in accordance with our beliefs and can hold seemingly incompatible views simultaneously. The diversity of experiences and beliefs motivating human actions, and the diversity of actions helping to shape beliefs, presents a considerable challenge to communities and societies; reasonable people can disagree. It is therefore important that different values are understood, respected and acknowledged or taken into account, even if they cannot be reconciled.

Any consideration of ethics should emphasize two things: it is complex and yet it is what we all do. Ethics' complexity is one of its benefits. We, and the world we live in, are complex and simplifying it does not always work. Morality has its roots in cooperation between social beings over resources and a willingness and capacity to look for shared solutions to conflicts (Midgley, 1993). It grows out of relationships with one another; we have duties to animals that are 'close' to us. 'Close' includes those we depend on, those we impact on, and those we have deprived of resources for our benefit.

Nowadays we can distinguish between two types of ethics. The first is our common morality, the rules we learn as infants, the social and legal rules we learn later in life and our professional codes of conduct and so forth; this is our common sense and tradition. The second is the philosophy of our common morality, the analysis and criticism of moral theories; we recognize that common sense and tradition can be improved upon by reflecting on their foundations in philosophy. Ethical theories include the most well-known components of our common morality, rights and consequences, but there are many different theories. Like the different sciences – for example, behaviour, immunology, reproduction, describing animal functions – each ethical theory provides different insights, but each also has its own shortcomings.

The most well-known theory, utilitarianism, is based on the likely consequences of an action; the benefits are balanced against the harms. For instance, humans have benefited from breeding poultry to produce many more eggs than their wild counterparts once did, providing a relatively cheap and plentiful source of food. This is taken to outweigh the birds losing their eggs and some freedoms. That does not, however, mean all harms to animals are outweighed by the benefits to people. As a society we accept that there are some things we cannot do to animals, no matter what the benefits. Currently, there is debate over the confinement of layer hens in small and barren cages. Rights or deontological theories set a limit to actions, regardless of the benefits. They are a social device that makes it easier for people to live with each other by providing a protection or constraint on treatment. Animal rights theory appears to have evolved from an 18th-century reaction to humans apparently having no obligation to animals or to their treatment. Not surprisingly, the movement resulted in almost complete consensus in the need for speedy killing of animals when slaughtering or in eradicating vermin, and in repudiating cruelty to animals. Animal rights can, and commonly does, refer to any call for the fair treatment and protection of animals. However, understanding animal rights is confounded by two things. Firstly, the comparison with human civil and political rights; and secondly, the belief that, as subjects of their lives, animals should not be used by humans for any purpose. Such extreme or minority views contrast with the more common view, that animals have a right to be treated humanely. Unfortunately, in many circles 'animal rights' has almost become a pejorative term often or commonly associated with non-mainstream beliefs, protestors, extremists and criminal actions.

Four moral principles, originating from our common morality, have been used in medical ethics to deal with aspects of health care, and in agriculture (Beauchamp and Childress, 1994; de Boer et al., 1995; Mepham, 2000; Mepham and Forbes, 1995). These principles are: beneficence (the obligation to provide benefits and balance benefits against risks), non-maleficence (the obligation to avoid causing harm), respect for autonomy (the obligation to respect the decision-making capacities of free-willed beings) and justice (an obligation to be fair in distributing benefits and risks). This approach claims that whatever your interests and background, the four principles can be applied. Furthermore, they are prima facie principles, and each is binding unless it conflicts with another, in which case you must choose between them. The principles do not aim to provide answers to problems, but to provide a common set of moral commitments, language and issues to be considered in particular cases, before coming to your own answer, using your preferred moral theory or other approach to choose between conflicting principles. To ensure all interest groups are considered, an ethical matrix can be developed whereby each of the principles is applied to animals, farmers, consumers, the environment and so forth.

While ethical theories are useful, they can be 'impoverished since they emphasize impartiality and abstractness ... at the expense of giving proper weight to special relationships' of the sort inherent in animal husbandry and caring (Anthony, 2003). Reciprocal and persistent relationships, forged between individuals and promoting the welfare of both parties, may give rise to different understandings of how animals should be treated, aspects that the sorts of ethical theories described above have difficulty incorporating. Most significantly, they allow, and require, special consideration to those with whom we are in relationships, such as farm animals, and those we desire to see flourish and be happy, as the ancient or domestic contract between man and animals holds: caring for animals by providing for their needs. The ethics of caring focuses on nurture and empathy, calling attention to the inequitable distribution of resources, lack of trust and agricultural production, and asking questions about 'our willingness (or lack thereof) to share the burdens of being in a relationship', even when they come with some expense (Anthony, 2003). And as farmers and farm animals are in a special relationship, so too are citizens and consumers along with the food supply chain. Caring, which begins from the foundation of relationships, requires providing those in charge of animals with the skills, confidence and resources to care for animals – promoting the skills of husbandry rather than the skills of following directions, and systems that acknowledge the complexity of agriculture and enable human–animal relationships to flourish (Curry, 2002). Quality of human life is, it is suggested, due to or determined by things like love and companionship, knowledge and understanding, being prepared to compromise, treating each other with fairness and respect, acknowledging our interdependence, finding solace in each other and enabling each other to flourish. Perhaps two of the most important of these are knowing and understanding each other and encouraging or enabling each other to flourish. What do these mean for the relationship between animals and humans?

Virtue ethics is based on the capacity cultivated by experience and training to have emotions that make you feel like doing good things and remind us of human limitations. For example, the capacity for reverence or to be in awe of, or have respect for, things outside of our control. A good person has such qualities and it forms the basis of actions rather than reverting to rules, isolation from emotions (feelings that motivate) impairing judgement. A chance to flourish (to live with dignity and have a life worth living) is a more adequate basis than rights and utility (Nussbaum, 2004).

Another ethical theory is pragmatism (McKenna and Light, 2004), which not only considers the consequences of an action, but also emphasizes the legitimate and necessary role that emotions and sympathy play in moral reflection and choice. Something is right if, based on the practical consequences, it works. Casuistry, or care-based reasoning, is informed by an intimate understanding of a particular situation and history. It is arguably most evident in the skills of stockmanship and animal care based on a respect for the essence of the animal.

The different ethical theories arise from a long tradition of reflection on different viewpoints, but a final example of ethical guidance is that based on the capabilities and nature of animals (and people): what enables them to flourish (Tulloch, 2011)? Some suggest continuing to live; being in good health and not disfigured; having opportunities to move, form attachments, play and even strive for goals; being treated with respect, dignity and fairness; and having some control over one's environment. For many animals, a good life also includes attachments to humans.

The scale of our dependence on, and compassion for, other creatures is a peculiar human trait resulting in significant bonds and the form or fairness of this relationship evolves in part through critical reflection. Ethics suggests that, in future, we may have to grapple with providing animals with a more dignified existence, one that allows them to flourish, or to have lives worth living, through, for example, being able to live longer, remain healthy, have freedom to interact with and experience the environment, establish relationships, play and have control over their lives. In apparent contrast, one account of ethics is that it is merely an attempt to fit reason to conventional beliefs, its use limited in difficult decisions where we must draw on other qualities such as intuition or even self-preservation (Gray, 2002). To add weight to this view, there are two ways to make moral decisions: fast and instinctive, or more emotional; and slow and deliberative, or more rational (Greene, 2013; Haidt, 2012). Our western focus on reason may not be the best approach.

The way we interpret or choose between conflicting viewpoints, the weight we give to different theories, and different interest groups, is determined by who we are and how we see ourselves within the world, the day-to-day influences that affect our reasoning, and the historical and cultural circumstances and context that provide an insight in determining rights and wrongs. The important ethical points are revealed by considering connected or related influences, in other words the 'story' or narrative of the issue. For example, choices made by an individual may differ from those made by that same individual within the context of his or her family or profession.

A myth is a story that arranges the past, whether real or imagined, in a way that reinforces our values and aspirations, and expresses our group or culture's shared world view. Myths encompass two ideals, namely: (1) notions based more on tradition or convenience than on fact, what we believe is real; and (2) the widely shared stories we tell about how society ought to be organized, what we want to be real (Browne et al., 1992; Fisher, 2002). Examples abound throughout history. For example, Darwin's theory of evolution undermined the idea that man was made in the image of God and he or she is a uniquely rational being. Thus, traditional moral values, based on the idea that human life has unique value above that of non-human life, have changed, resulting in the increase in the value of non-human life. Similarly, since it is linked to the very origins of civilization, and it provides us with some of our most basic needs, farming, according to the agrarian myth, is especially virtuous. The myth informs us that a good farmer understands and works with nature to feed a hungry world. Since these values are held in esteem by society, it follows, some argue, that society should adopt policy favourable to agriculture. Furthermore, a successful farmer is one who works hard, producing as much food as possible (an abundant harvest is a sign of heavenly grace), giving rise to the productionist narrative, the level of production being the sole measure of the success of farming. Therefore, instead of determining appropriate treatment of animals on the basis of ethical theories (rights, consequences, justice etc.), superior animal growth and reproductive rates are an indicator of the right way to farm. It is this contradiction, between maximizing productivity (farming as business) and farming in harmony with nature (farming as a way of life), which led farmers to value production over feelings and emotions.

Animal agriculture is often depicted in various images (e.g. the barnyard hen and the grassland flock) and beliefs (e.g. farming is the backbone of the economy). Traditional portrayals are being challenged by 'new perceptions' (Table 4.6), necessitating the integration of knowledgeable research and analysis with the genuine concerns of individuals and advocacy groups (Fraser, 2001; Pielke, 2007; Thompson, 1999a).

While myths help shape the way we see the world and guide our activities, they can also misinform us when based on untrue or outdated assumptions. In *Sacred Cows and Hot Potatoes*, Browne et al. (1992) recognize and acknowledge how our views of the world are based on these narratives and myths, making for a better understanding of the use of animals in farming. Among the perhaps outdated myths, at least within the US, were: confusing farming with the countryside, basing decisions on the 'average' family farm, confusing production with productivity, confusing farm prices with farmers' income, forgetting that agriculture depends on the world

Table 4.6 *Contrasting traditional and new portrayals contributing to a propaganda battle as defenders and opponents oversimplify animal agriculture for their own interests (Fraser, 2001).*

Traditional portrayals of farming	New perceptions of farming
Beneficial for animal welfare	Detrimental to animal welfare
Mainly controlled by families and individuals	Mainly controlled by large corporations
Motivated by traditional animal care values that lead to profit	Motivated by profit
Augmenting world food supplies	Causing increasing world hunger
Producing safe, nutritious food	Producing unhealthy food
Not harmful, and often beneficial, to the environment	Harmful to the environment

economy, equating good farming with a healthy environment, assuming farm programmes are food programmes, and assuming that a government programme will do what it says. Finally, one of animal welfare's myths is that consumers are driving demand for welfare. While they, or at least food retailers, might be, so too many non-consumers (e.g. Ruth Harrison was a vegetarian) have driven, and continue to drive, animal welfare. Narrative influences, then, try to make sense of our experiences, traditions and culture. In a sense they tend to justify differences in moral behaviour, whereas ethics is often regarded as universal. Myth-making is about our choice of future pathways. The pastoral myth associates shepherding and purity and virtue, helping to blind us to anything harmful.

Reducing consideration of the treatment of animals in farming to one or more dominant theories of ethics is like understanding animals (or humans) by understanding their immune system, or their reproductive system. It neglects both the richness of ethics and the complexity of the relationship between parties, just as immunology and reproduction neglect the cardiovascular system and the gastrointestinal function. The importance of farming to humans, the importance of species-specific characteristics to animals and the acknowledged place of animal husbandry all risk being neglected (Thompson, 2007).

We might conclude this section by asking if economics tells us what is good, or is there a danger the ethics will be taken over by economics? Is materialism the enemy of a good society given humane ideals are happiness, creation of beauty, justice, pursuit of truth, tolerance of or delight in diversity and the perfection of human nature (Flynn, 2012)?

Being human means balancing our ways of seeing the world, whether through our science, emotions, morality or myths. Aaltola (2005) emphasizes theory, pluralism and particularities of specific situations. Maybe we are heading towards the special importance of context rather than the universal ethics of animal use. We cannot be ethical about something we do not know. If we only know the price, the weight and the colour of our meat then that is all we can base our choices on. It is a poor substitute for seeing, feeling, understanding, respecting, admiring, trusting or having faith in animals and farming. While animals and farming reflect our ethics, economics dominates choices of product attributes. It is to economics we now turn since one of the main drivers of intensification is financial. We cannot simplify ethics by forgetting economics and we cannot let economics decide the future of humanity's relationship with animals.

Economics is about our behaviour in pursuing our own interests or preferences, the trade-offs we make between our preferences for food and an animal's discomfort in producing it (Bennett, 1997; Christensen et al., 2012; Lusk and Norwood, 2011b; McInerney, 2004; Norwood and Lusk,

2011). We cannot have food production without impacting on animals and through economics we balance the use of animals with the discomfort in doing so. How we decide what we want, how to best use our resources, who benefits from them and who pays for them are all expressed through economics. It is not just about money or finances, though money is a convenient measure of our preferences.

Animals are a resource like anything else in farming, except that they also have their own interests. What people want determines an animal's value, suffering a regrettable side effect except that it is also a 'good' desired by humans, whether they consume animal products or not. Animal welfare is determined at a societal level and not just by individual farmers, since the wider community cares how animals are treated. It is therefore important to understand people's perspectives, be they informed by animal behaviour, science, shepherds and cowboys, anthropomorphism, past experiences, ethics, cultures, myths, prejudices, what alternatives are available and so forth. It is this inescapable trade-off between animal and human welfare that is explored by economics.

The economic ethic (cost–benefit analysis) is based on the advantages and disadvantages to society of which money is the common unit in the marketplace. However, there are problems as the market does not reflect the true values of society but only those things exchanged through markets. In addition, the market value can be distorted by factors like subsidies and taxes and does not reflect the true value as the preferences of people not involved in the market are not considered. Finally, those involved in the market, for example, retailers and consumers, are not necessarily fully informed of things like animal welfare. Unless externalities like animal welfare are included in the market, animals, like any resource, can be exploited and animal welfare lowered. Governments are thus usually involved to ensure efficiency and fairness (e.g. setting minimum standards and redistributing the costs and benefits).

Consumers vary in the value they put on products as choices can be constrained by income and dependent on prices. They may also seek products with the most preferred overall collection of characteristics (animal welfare is important but not all-important). For whatever reason, consumers may not want to support higher animal welfare systems. For example, one study asked how much people were prepared to pay to improve the welfare of 1000 hens. The study used a novel technique – people had to bid to move birds from a caged system to an aviary with free access to a range and make decisions based on spending their real money. A third of people were unwilling to pay anything and most would only pay a paltry amount. A minority, less than 10%, were willing to pay a significant amount (Lusk and Norwood, 2012). And any additional costs imposed on the consumer may alter their purchasing behaviour, either preferring cheaper products or seeking higher welfare products providing they can be identified and trusted.

This necessarily brief introduction to economics indicates there are several points worth noting about animal welfare. Firstly, an increase in food prices is not necessarily a bad thing (many vary seasonally, with changes in taxation, and with in-store specials). What effect does a change in prices have on consumer behaviour? It seems to be an accepted belief within animal welfare circles that the impact of changes in practices should be balanced against the impact on food prices, especially for those on lower incomes depending on lower-priced and hence intensively produced foods. While not disregarding the importance of food prices for low-income families, this would not seem to justify lower prices for all (there may be other mechanisms for supporting low-income families, for example). While some changes to animal welfare are not costly, others, especially those involving large capital-intensive systems, undoubtedly are. It is necessary then to estimate the proportion of the increase in production costs within the final price. Furthermore, the proportion will vary depending on if the product is 'raw' such as a table egg, or 'processed' such as a breakfast egg at a restaurant. Food supply chains are reasonably complex, and we do not yet know how equitable various parts are in sharing the costs of improvements to animal welfare in practice. It is important to know the costs of various improvements and how they might be distributed across the food supply chain.

Secondly, there is no reward in the marketplace for high standards of animal welfare (the majority of animal welfare improvements do not result in economic advantages for the farmer, for example; Gocsik et al., 2015). Animal welfare is determined by good animal husbandry and stockmanship, the value the farmer places on it. In other words, animal welfare is a value attributable to farmers, not consumers, and farmers need economic incentives if society wants higher standards. In contrast, the 'market' can be perverse in paying less, such as the cost-price squeeze felt by dairy farmers in the UK when the major retailers decided to pay them less for milk (Figure 4.4).

Finally, higher animal welfare standards have a cost to system productivity (Figure 4.5). Measuring the benefits of improvements in animal welfare is difficult. Many studies have assessed consumers' willingness to pay for food produced in higher animal welfare systems but, as noted, there are problems with using this measure. People tend to respond to surveys as citizens, differently to how they shop as consumers, and animal welfare is not a private good captured by market transactions, but the value society attaches. Our valuations also emerge with

Figure 4.4 A UK dairy farmer protesting supermarkets reducing the payment for milk from 20p to 14p per litre of milk in 2002 while consumer prices remained at 45p (*Photo: Adrian Sherratt/ Alamy*). The sentiments were not lost on the supermarkets, this one demonstrating its 'commitment' to dairy farmers in 2016 (*Photo: Mark Fisher*).

experience and learning. Finally, the links between consumers and farmers are anything but simple; there are many opportunities for people to exploit, distorting the connection necessary for the market to accurately inform production.

There are many potential economic means of modifying animal welfare. They include:

- taxes and subsidies (some countries base production subsidies on successful animal welfare inspections);
- labelling to enable consumers to support particular systems (depends on acquiring the information from the label to make a reasoned and informed decision);
- tradeable permits whereby those in more animal welfare friendly systems can effectively subsidize those in less friendly systems;

- quotas to ensure production is limited and thus farm returns appropriate to cover the expenses of favourable systems;
- premiums for favourable, animal welfare-enhanced products;
- consumer subsidies;
- provision of information and education to facilitate changes; and
- farm assurance highlighting advantageous systems.

None of these market mechanisms are expected to provide a single solution and state intervention is also required. And there are undoubtedly other novel market mechanisms as economic understanding may help share the burden. There are limits to improving animal welfare through either the law or the market, and enhancing animal welfare requires both.

While the contribution of economics to animal welfare is yet to be fully realized, it is perhaps best summed up by Harrison (1971): 'We must as a

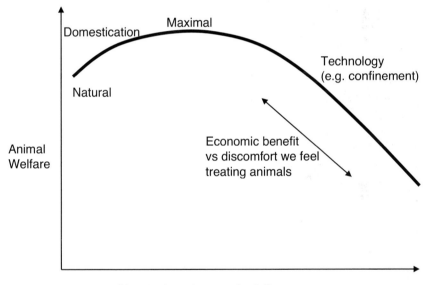

Figure 4.5 The hypothetical relationship between animal welfare, the vertical axis, and the productivity of the animal production system, the horizontal axis (from McInerney, 2004). In a wild or natural population of animals, with no human management or control, animal welfare might be imagined as natural. With domestication, animals benefit from food, shelter and protection from predators in return for production, which increases until a point when welfare is maximal or optimal. However, some practices, such as close confinement, enable further increases in productivity though imposing welfare costs on the animal. Eventually productivity reaches a limit beyond which the animal is unable to cope. The welfare of animals is determined by individuals and society reaching an acceptable balance between the benefits of animal use and the discomfort felt about the conditions animals are exposed to.

nation decide which systems are biologically and ethically acceptable and then make sure that the farmer is adequately recompensed for using these humane systems.'

Can society deliver good animal welfare?

So now we have come to understand animal welfare, rightly or wrongly, not just as the state of the animal and the skills, expertise or stockmanship of those caring for them, but as a complex interaction between the animal/farmer and many of the actors in this chapter, from individuals like Ruth Harrison to super-state organizations like the World Organisation for Animal Health (OIE).

Farmers, and several others, like transport and saleyard operators, and slaughtermen, provide the care for animals. But they also have to 'listen' to, or judge, all the guidelines, rules, advice and information that society provides them with, and interpret them within the context of their own values, circumstances and communities. The diversity of societal influences in this chapter, from religious tenets to media exposés, makes that anything but a simple exercise. Especially when those individuals are at the sharp end of balancing animal compromises against human benefit, costs and benefits also gained and borne by many others in society.

Animal husbandry's ethical foundation is that it is appropriate to use animals as long as that use is humane. But modern humanity has lost the hunter's recognition of the individual and is becoming increasingly estranged from the natural world (Comstock, 2000). Farmers' skills of stockmanship and business acumen, and their personal values and goals, provide much of the impetus for how animals fare. However, and reflecting the contentious nature of animal welfare, there are also a myriad of state, market and private initiatives and expectations: a veritable animal welfare industry. Is this 'industry', too, disconnected and complex? Will it improve or further impact on farming? Perhaps it is best positioned to deal with some of the simple causes of poor welfare but is it able to deal with the financial and underlying institutional drivers of intensification contributing to poor animal welfare? Animal welfare might be able to be improved by finding ways to support and sustain the forms of farming that are considered acceptable, rather than telling people what to do, including prohibiting or limiting practices.

Animal welfare is a social construct, evident in the social initiatives involved (e.g. legislation and codes, animal advocacy interests, transnational governance, business opportunities). Accepting that the 'animal welfare industry' is necessary, have its efforts significantly enhanced animal welfare or are they valuable in their role as an antidote to our loss of contact with farm animals (and nature)? Are they relieving the guilt of 'decadent barbarians' (J. Armstrong, 2009), disconnected modern humans? If this latter possibility is important, are we focussing wrongly on the farmer? What this chapter has shown is that many in society are involved in animal welfare but that the overriding characteristic of that involvement is in telling farmers what to do, often berating them for being greedy profiteers when most people have not thought deeply about the wider implications of doing so. As McInerney (2013) notes, we are in effect 'behaving like spectators at a football match, shouting at players and the referee to tell them what they should be doing, whereas we weren't the ones who were playing the game'. But, and like the game, animals are important to us all and this is the subject of Chapter 5. Sadly, those responsible for animals feel increasingly as if those beyond the farm gate are not listening but merely telling them what to do. We must live within our means, so disciplines like sociology are as important to animal welfare as are the perspectives of those who know animals, be they farmers, veterinarians or animal advocates. Only by focussing on and enabling those responsible for the care of animals and the husbandry skills (not necessarily the skills of following society's diverse and varied directions) can we make progress. We must also reward the farming systems that promote the human–animal relationships that society values, while dealing with the tension between what society advocates for animal welfare and what those in charge of animals can realistically deliver within the constraints of, for example, training, climatic events, economics and technology.

The seemingly complex expectations and influences provide a challenge for those charged with reaching a societal consensus. However, although there are many seemingly divergent values, they probably coalesce into a few themes focussed upon, for example, a fair relationship between animals and people. Resolving controversies characterized by different values may require novel approaches, such as reframing problems, providing new types of facts, and commitments to principles of engagement, for example, demonstrating that different views are understood, valued, respected and considered even if they cannot be reconciled. The challenge is to find a way to negotiate our way as a community through so many different values, where there is no one line of argument that deals with all situations we encounter in our relationships with animals with the associated inevitable and unavoidable tensions and inconsistencies. The issues are the domain of ethics, the systematic reflection of moral issues in the public sphere, including articulating those tensions, reducing as much as possible their impacts, and making it clear where value judgements are required. The tasks are both intellectual and social: empowering the full range of voices and encouraging shared understanding with the intent of reaching some consensus.

Combining science and ethics, sciences and humanities, is not new. Aristotle pointed out over 2000 years ago that acting with a rationality linked to an understanding of values was essential for the well-being of individuals, communities and societies (Heap, 1995). He described three intellectual virtues: scientific knowledge, technology and prudence or action about what is good or bad. Subsequent history saw western intellectual life divide into two cultures: the arts or humanities and the sciences (Snow, 1959). A colleague once remarked, before the 1970s animal welfare used to be about people, now science has hijacked it. Increasingly, ethics and sciences are now coming back together in animal welfare.

Collectively, the individuals and organizations highlighted in this chapter, although all generally working towards a common goal, the welfare of animals, have the makings of an animal welfare industry (society is producing guidelines to ensure farmers respond to society's demands for being more responsible!). It is huge, multifaceted and complex, presenting those in charge of animals with a bewildering array of advice, demands and expectations. This industry of animal welfare 'middlemen' is keen on developing a fix that inevitably costs more, or at least makes animals, farmers and the environment work harder. The challenge, to which we return in Chapter 6, is to bring it all together in a way that ensures animal welfare is sustainable, to question which parts of this 'animal welfare industry' are changing the well-being of animals for the better, and which parts are making humans feel better in the short term but contributing to the treadmill? However, before doing so, it is necessary to reflect on the breadth and nature of our relationship with farm animals, the subject of the next chapter.

Further reading

Report of the Committee to Consider the Ethical Implications of Emerging Technologies in the Breeding of Farm Animals. **Michael Banner, G. Bulfield, S. Clark, L. Gormally, P. Hignett, H. Kimbell, C. Milburn and J. Moffitt.** HMSO, London, UK, 1995.

Improving Farm Animal Welfare. Science and Society Working Together: The Welfare Quality Approach. **Harry Blokhuis, Mara Miele, Isabelle Veisser and Bryan Jones**. Wageningen Academic, Wageningen, the Netherlands, 2013.

The Evolution of Morality and Religion. **Donald Broom**. Cambridge University Press, Cambridge, UK, 2003.

Understanding Animal Welfare. The Science in its Cultural Context. **David Fraser**. Wiley-Blackwell, Oxford, UK, 2008.

Livestock Ethics. Respect, and Our Duty of Care for Farm Animals. **Gordon Gatward**. Chalcombe, Lincoln, UK, 2001.

Physiology and Behaviour of Animal Suffering. **Neville Gregory**. Blackwell Science, Oxford, UK, 2004.

Ethics. A Pluralistic Approach to Moral Theory. **Lawrence Hinman**. Harcourt Brace, Fort Worth, TX, 1998.

Livestock, Ethics and Quality of Life. **John Hodges and In Han**. CAB International, Wallingford, UK, 2000.

Animal Welfare, Economics and Policy. **John McInerney**. Defra, London, UK, 2004.

Animal Pragmatism. Rethinking Human-Nonhuman Relationships. **Erin McKenna and Andrew Light**. Indiana University, Bloomington, IN, 2004.

Animals and Why They Matter. **Mary Midgley**. University of Georgia Press, Athens, GA, 1983.

Awe for the Tiger, Love for the Lamb. A Chronicle of Sensibility to Animals. **Rod Preece**. Routledge, New York, NY, 2002.

The Evolution of Everything. How New Ideas Emerge. **Matt Ridley**. Fourth Estate, London, UK, 2015.

Animal Rights & Human Morality. **Bernard Rollin**. Prometheus Books (Revised edn), Buffalo, NY, 1992.

Farm Animal Welfare. Social, Bioethical, and Research Issues. **Bernard Rollin**. Iowa State University Press, Ames, IA, 1995.

Animal Revolution. Changing Attitudes towards Speciesism. **Richard Ryder**. Berg, Oxford, UK, 2000.

The Ethics of Intensification. Agricultural Development and Cultural Change. **Paul Thompson**. Springer, Dordrecht, the Netherlands, 2008.

Veterinary & Animal Ethics. Proceedings of the First International Conference on Veterinary and Animal Ethics, September 2011. **Christopher Wathes, Sandra Corr, Stephen May, Steven McCulloch and Martin Whiting**. Wiley Blackwell, Oxford, UK, 2013.

CHAPTER FIVE

People are people through animals

There is an African saying, *Muthu ndi muthu nga numwe*, meaning 'a person is a person through another person'. In other words, the relationships we have with other people define our lives (Kasene, 1994). In this chapter, the relationship between animals and people is examined, suggesting that humans also exist in terms of the relationships we have with animals. A characteristic of human nature is that we do not naturally exist in species isolation, but draw in, domesticate and live with a great variety of animals (Midgley, 1983). Furthermore, a strong interaction with animals, indeed with all life forms, is critical to the health and well-being of people (Bustad, 1988). 'Animals are our food, they are our thoughts' is a saying attributed to Navajo hunters (Nelson, 1993). Other cultures have similar beliefs, for example, John Donne's (1572–1631) 'No man is an island.' Such is our interdependence, perhaps people are also people through animals, and animals are animals through people. Compromising animals for human good, for legitimate societal interests, is wide-ranging.

Animals play an important part in our lives, reflecting the long, complex, interdependent, enduring, diverse, broad and overlapping relationships between them and us (Benton, 1993). They are not just simply our companions, sources of food or pests, animals are us, and there are many ways to understand the multitude and complexity of relationships. For example, categorizing them as farm animals, horses and pets, wild animals and animals in science (Webster, 1994). In contrast, a fabled Chinese encyclopaedia, the *Celestial Emporium of Benevolent Knowledge*, categorized animals as belonging to the emperor, embalmed, tame, sucking pigs, sirens, fabulous, stray dogs, included in the present classification, frenzied, innumerable, drawn with a very fine camelhair brush, having just broken the water pitcher, or those that from a long way off look like flies (see Harford, 2016).

Although there are different ways to think of the relationships, one (Table 5.1) is particularly valued for its broadness and acknowledgement of overlapping categories. As part of a review of human rights (taking the

Table 5.1 *The diversity of relationships between humans and animals (Benton, 1993).*

Source of labour
Use of animals' strength, stamina, obedience and psychological reliability to help or replace human labour
- mules, camels, donkeys, horses, llamas and cattle
- sniffer, guide, and sheep and cattle dogs; hunting dogs

Meet humans' and animals' organic needs
Animals and animal products as a supply of food, fibre, tools, weapons, testing
- meat, milk, eggs, fertilizers, animal feeds
- skins and pelts, feathers, wool and furs for clothing and shelter
- bones, teeth and tusks for needles, harpoons, awls, forks, throwing sticks, knives and ornaments
- studying human and animal diseases, testing safety (e.g. canary in a coal mine) and efficacy of medicines and products

A source of entertainment
- zoos, circuses, nature reserves, wildlife books and films
- horse and greyhound racing, riding, hunting and fishing
- breeding, showing and agility testing, bull-, dog- and cock-fighting

Provide moral and intellectual learning
Observations, science and the keeping of animals provide opportunities for drawing similarities between humans and animals
- wildlife films, zoos, field observations of animal behaviour and ecology, pets

Commerce, wealth, power and prestige
A significant resource for economic systems and market economies
- farming for food and fibre, animals in work and being used for entertainment and in education, research and safety testing

Table 5.1 (continued).

Coercive maintenance of social order and the protection of property
The capacity of animals to threaten and harm humans in antagonistic human relations
- guard, police and sniffer dogs, police horses

Companions or pets
Household or family members generally entering into reciprocal affective bonds and communicative interactions providing emotional support and companionship
- cats, dogs, birds, fish, frogs, snakes, turtles, lions, tigers

Animals as means of thought or symbols
Means of expressing good and evil, emotions, character and relations, natural forces and processes, objects of worship, structural units of human societies by using the attributes of animals for metaphorical and symbolic convenience to teach us about ourselves
- the bat out of hell, the filthy pig, Aesop's fables, big bad wolf

Wildness
The symbolic ideas of freedom and of being unrestrained by civilization; natural, wild, indigenous, non-domesticated and feral species; potential sources of food, sport and entertainment, traditional medicines and competition (e.g. deer, pigs); animals adapted to and associated with humans (e.g. rats, mice, sparrows, pigeons). The spread of humans and associated ecological impacts, and the right to interfere with or to protect habitats mean wild animals are also part of human relations and institutions. Although wild, they are affected by, and affect, humans and are thus a significant part of our world.

moral status of individuals as a basis for reviewing human rights), Benton (1993) in *Natural Relations. Ecology, Animal Rights and Social Justice* considered the features of, and the part played by, animals in human social life, emphasizing the range of human needs and wants which animals meet. Benton goes on to note that the extent of these relationships is remarkable; animals are part of our lives and humans cannot live without animals, or at least not without having an impact on them. And people affect animals in different ways, for example, through disturbing ecological systems and habitats, introducing invasive species and causing pollution; altering environments through crop production and night-time lighting; keeping them for farming; and directly harming them by hunting, toxicity testing and slaughtering (Fraser and MacRae, 2011). Whether knowingly or unknowingly, directly or indirectly, animals are part of our lives, our histories, our past, present and future, our characters and our very essence.

Animals are used as a source of labour by utilizing their muscle strength, stamina, obedience and psychological reliability. Some examples are well known, such as oxen, horses, donkeys, buffalo, camels and elephants used as draught animals in many parts of the world and throughout history. Animals also work for humans in more complex and sophisticated ways. For example, sheepdogs learn to coordinate their activities with humans by interpreting auditory and behavioural cues, as well as act autonomously in undertaking those actions (Figure 5.1).

Figure 5.1   *Gin*, a working sheepdog, herding poultry (*Photo: Mark Fisher*).

Animals also provide labour in more unusual ways, one of the more exotic being the civet cat living in forests and on farmlands from the Himalayas and China to the Philippines and Indonesia. An omnivore, it feeds on small vertebrates, insects, ripe fruit, seeds, palm sap and coffee cherries. Coffee beans excreted by the cats are used to produce *kopi luwak* (civet coffee), a rare and very expensive coffee. The beans are pitted by the cats' digestive enzymes giving the coffee a distinctive gamey flavour, pleasantly sweet without bitterness, and syrupy, chocolatey, earthy and musty with jungle overtones. Bird's nest soup (made by small birds with their saliva), argan oil (a cooking oil made from nuts of the argan tree recovered from the faeces of Moroccan goats) and honey are examples of animals' digestive systems providing or at least processing food for humans.

Food is one of the primary reasons for hunting and for farming animals, providing meat, milk, eggs and other products. Animals also provide clothing and accessories (e.g. wool, fur), shelter (e.g. hides, skins), tools, weapons and ornaments or jewellery (e.g. bones, shells) and fertilizer (e.g. blood and bone, guano). The cat, once sacred in Egypt, was embalmed and buried in such great numbers that their mummified remains were once excavated and shipped to England at the beginning of the 20th century to be ground up for agricultural fertilizer (Benton, 1993).

Some discourses on animals tend to emphasize farming as providing meat, the answer being to become a vegetarian. While a major product of modern farming is certainly meat, along with milk, eggs and so forth, meat is only part of the story. In 2001, in response to the impacts of the possibility of devastating diseases affecting the beef industry, the innumerable uses of beef by-products were summarized (Klinenborg and Modica, 2001). It began by identifying the obvious ones such as meat, leather and hide, processed animal feed, pet food, and garden supplies such as blood and bone fertilizer, but identified many other uses for rendered cow parts (Table 5.2). Some uses are quite unusual, for example, catalase recovered from the liver for use in contact-lens care products, blood was a component of plywood adhesives, collagen a binder for flammable substances in the manufacture of matches and to improve the 'crispness' of banknotes, and gelatin was used in cooked meats to improve slicing.

The extent of the use of cattle products, evident in these examples, is a modern equivalent of the well-known saying, every bit of a slaughtered pig was used 'except the squeal', reflecting the determination that nothing grown or fed from the land would be wasted.

Servicing human needs with animals and their products, has enabled many people to make their economic livelihoods (Figure 5.2). Animals are resources in commerce whether used for work, food, entertainment or learning. We farm fish, undertake research on monkeys, test chemicals on

Beannachdan Air Na Cu Caorach (blessings on the sheepdogs) at Lake Tekapo, New Zealand (*Photo: Mark Fisher*).

Two examples of how inseparable dogs' and humans' worlds can be. A young autistic boy in hospital with his ever-present assistance dog (*Photo courtesy Louise Goossens / Capital and Coast District Health Board*) and another youngster and his dog, surfing (*Photo courtesy Heather O'Brien*).

A modern farmed red deer hind (*Photo: Mark Fisher*). Note that this hind is heavily pregnant, in this instance with a hybrid Canadian wapiti x red calf, part of work to find the most efficient system of production, nominally a large male mated to a small female. Note also that the pasture in the foreground has gone to seed. Deer, being browsing species, calve in early summer after the spring peak of farm pasture production, necessitating research to induce earlier mating and thus calving, and thus more efficient use of pasture for lactation.

Sheep bred by Robert Bakewell as portrayed in John Ferneley's 1823 painting of 'Sir John Palmer on his favourite mare with his shepherd, John Green, and his prize Leicester Longwool sheep' (*Courtesy Leicestershire County Council Museums Service*). To satisfy the demands of the growing human population of the Industrial Revolution, Bakewell bred fat animals that matured earlier.

Part of an advertisement for a veterinary pharmaceutical to treat and control internal and external parasites (*Photo courtesy Merial NZ, now Boehringer Ingelheim Animal Health New Zealand*).

Examples of extensive and intensive or landless farming (*Photos: Mark Fisher*). Globally most protein is produced intensively with smaller amounts produced on grassland systems. All systems have inherent efficiencies and inefficiencies, yet their products are marketed and consumed similarly, making them susceptible to similar market forces.

Riding on the dairy cow's back (*Courtesy John Ditchburn*).

Individuals in farmyard settings know each other whereas few of us even see, let alone know, the depersonalized industrially farmed animals such as those in barns where it is difficult to even see all the birds (*Photos: Mark Fisher*). A stocking rate of seven birds per square metre led Ruth Harrison (1991) to comment that barns were 'hardly an improvement on battery cages'.

Saint Jerome Extracting a Thorn from a Lion's Paw by Master of the Murano Gradual (d. 1473) from the second quarter of the 15th century (*The J. Paul Getty Museum*, Los Angeles, courtesy of the Getty's Open Content Program).

The role of people in animal welfare: the next revolution? While farmers (*Photo: Ross Buscke*) are directly responsible for the animals (*Photo: Mark Fisher*) and are the most important determinant of their welfare, people (*Photo: European Commission*) also have an indirect role and thus responsibility.

City with Animals (Stadt mit Tieren) by Max Ernst (*courtesy Solomon R. Guggenheim Museum, New York, Estate of Karl Nierendorf*).

The Trolley Hunters (courtesy Banksy).

Bully Todd and a young bull whisperer (*Photo courtesy Fiona Fisher*).

Table 5.2 *A selection of some of the uses of rendered bovine products (from Klinenborg and Modica, 2001).*

Organs, glands and limbs
Lungs, adrenals, pancreas, placenta, dura mater, heart, trachea, tendons, nasal septum and mucosa, uterus, intestines, umbilical cord, used in cosmetics, drugs and pharmaceuticals including steroids, insulin, surgical implants and patches   Gall, bile, foetal calf serum, liver, used in laboratories, industrial processes and as cleaning agents   Intestines used for sutures, musical strings and racquet strings   Feet and hooves used in glue, buttons, handles, lubricants, soaps and fire extinguisher foam
Blood
Serum, proteins, nutrients, hormones, factors and other compounds used in laboratory systems such as in vitro growth of cells and vaccine production
Tallows (fat derived from meat, bone, hooves and horns)
Edible tallow used in shortening for baking, in combination with vegetable oils for frying foods, and in chewing gum; inedible fats and oils in industrial tallows   Fatty acids used in plastics, tyres, candles, crayons, lubricants, soaps, fabric softeners, pesticides and herbicides, dispersing agents, synthetic motor oil, gel cultures, cosmetics and pharmaceuticals, explosives, waterproofing agents, rubber, textiles, corrosion inhibitors, emulsifiers, coating agents and detergents   Glycerine used in medicines as diverse as cough syrups, tranquillizers, antibiotics, contraceptives, ear drops, topical analgesics and antibiotics, suppositories; as a sweetener and in dynamite, cosmetics, liquid soaps, candy and liqueurs, inks, lubricants, antifreeze, toiletries, sunscreens, hair and skin products, veterinary pharmaceuticals, agricultural chemicals, cleaning and polishing products
Connective tissues and skins
Collagen used in surgical and medical products, research reagents; gelatin for use in food including confectionery, desserts and dietetic products (dietary breads, biscuits, powdered soups); cosmetics, industrial and photographic products

mice, display pandas, make handbags from crocodiles, race horses, fight bulls and hunt deer for many reasons, important among them our livelihoods and careers.

Animals are also part of human sports, recreation and entertainment,

Figure 5.2 Sheep being traded at a saleyard in New Zealand (*Photo: Keith Fisher*).

though used in different ways. Firstly, bull- and cock-fighting, bear- and badger-baiting are events condemned by some individuals and communities. Secondly, many people find recreation in hunting animals such as deer, foxes, hares and pigs, often using dogs to find and capture them. While some of these hunting activities provide for human needs (e.g. meat, fish), many are also a sport, some arguably unsporting (Scully, 2002). Fishing also has important recreational and sporting aspects. Thirdly, horse and dog racing and the cowboy rodeo entertain those participating, whether as owners, spectators or gamblers. Fourthly, circuses, zoos and wildlife and farmyard parks enable more passive interaction with animals, at least in modern times. Circuses began as wondrous exhibitions of exotic animal species and chariot racing in ancient Egypt and Greece. The modern circus came about through a British cavalryman performing equestrian feats interspersed with clown antics, and later including tumbling, rope-dancing and juggling acts, as well as performing animals. Exotic animals were added to some circuses in the early 19th century. As well as being entertained, we also learn from animals in zoos, nature reserves and wildlife books, films and television programmes. Observations of animals in the wild teach us about their behaviour and ecology, and pets can help us to recognize and acknowledge a responsibility for the needs of animals.

Humans also expend significant effort reducing or minimizing the effects of other animals competing for the resources we share. There are many pests, mislocated invasive or nuisance animals including rats, mice, pigeons, possums, foxes, camels, coyotes, rabbits and badgers. Some are vectors and reservoirs of disease (e.g. bubonic plague, rabies and bovine tuberculosis), others predators of valued plants and animals, often crops and livestock. Ironically, indigenous species threatened by introduced or exotic species are afforded exceptional protection and public sentiment. Finally, some animals may directly threaten human safety (e.g. crocodiles, sharks, tigers, honey badgers).

Animals are used in art, entertaining us and perhaps helping us see and understand the world in different ways. For example, the sheep fed out in a pattern for airline passengers to view, and the pigs' ears depicting genocide (Figure 5.3). There are many examples of animals depicted in more traditional art forms.

In contrast to using animals as resources, cats, dogs, goldfish, parrots, snakes, birds and so forth can, as pets or companions, have a special relationship with people. As members of households some have become partners in quasi-personal or familial relationships communicating and contributing to affective relationships. They may act as child surrogates or contribute to children's development, highlighting the sophistication and quality, companionship and emotional ties that people and animals, especially cats and dogs, have. They may contribute to physical and mental health through being sources of unconditional love, living beings in a relationship with humans. The objects of play and laughter, animals help regulate our lives and act as a source of tactile stimulation, security and exercise, from keeping us warm in bed to keeping us company while dying (Dosa, 2007). These benefits are not only evident in our homes, but also in situations humans find challenging: prisons and nursing homes, or when suffering from various disabilities like polio, or being learning-impaired. Animal interactions can benefit people in varying states and conditions without necessarily compromising the animal (Bustad, 1988).

Long before they assumed their predominant roles as meat and milk, companions, sources of entertainment etc., animals entered the human imagination as messengers and promises, symbols and metaphors – anthropomorphism a measure of our proximity (Berger, 1980). Animals are a means of thought; because they are interesting we turn to them for symbolic expressions of our emotions, characters, relationships, natural forces and processes, and they are held sacred, protected, cherished, even viewed as ancestors and worshipped as gods. For example, the honey bee is symbolic of fertility and abundance, the pig of filth and obstinacy, and the bat of death and the underworld (Lawrence, 1993). The lamb is a symbol

Figure 5.3 Sheep fed a trail of silage to show support for the Otago rugby team (*Photo: Otago Daily Times*) and 'Forbidden Fleet' pigs' ears preserved in shellac beeswax and resin as part of an exhibition for people who may be missing part of their family by Angela Singer (*Photo courtesy New Zealand Herald*).

of purity, self-sacrifice and innocence, the coyote a trickster, the wolf of cruelty, ferocity, cunning and evil (although also a symbol of the wild), and the owl of wisdom, death and prophecy. Like myth and folklore, animal symbols help us to make sense of the world by expressing 'truths' that are deeper than that which we can observe. We choose certain characteristics, for example, bees represent the spirit of cooperation, intelligence, social solidarity and industrious behaviour, and ignore others, for example, bees' killing of unwanted hive members. These representations may vary between human cultures; in the Old Testament, for example, bees are aggressive. Perhaps the ultimate animal symbolism is that animals are, or can be, wild and free, independent of domestication (e.g. wolves, whales). In other words, unrestrained by the burdens of civilization, unaffected by the degeneracy of domestication and a regulated existence, and beyond human social power and domination. Many have no history of a relationship with man other than being potential prey, a competitor for resources or the source of curiosity. Some animals are iconic, such as wolves, deer and pandas. Some, for example, urban foxes, rats and sparrows, no less wild, have adapted to human conditions, while others have a more fragile and marginal existence as humans transform their environments. Animals give representation to nationalities (e.g. kiwi – New Zealand; bald eagle – US; lion – UK; rooster – France; eagle – Germany; bear – Russia) and especially modern sports teams (e.g. cubs, lions, panthers, wolves, brumbies, bears, sharks). Our language is rich with animals in:

- proverbs: milk the cow but do not pull off the udder; when the fox preaches look to your geese; every animal knows more than you do; it is not the horse that draws the cart but the oats;
- metaphors: busy as a beaver; cunning as a fox; straight from the horse's mouth; a fly in the ointment; and
- clichés: cry wolf; the black sheep; blind as a bat; flogging a dead horse.

Over 30,000 years after animals were portrayed in the cave paintings of Stone Age times, animals are still part of our cultural activities. The images portrayed in animated children's television pictures like *Cow and Chicken* and *Catdog*, each a pair of siblings, the former with human parents, the latter conjoined, are perhaps as puzzling yet as fascinating as the cave paintings and myths.

The diversity is also remarkable, not just in the breadth of the relationships and uses but also through history. In different regions and in different societies, different animals have been an integral part of humanity. The relationship is long and enduring, and the change from idyllic, harmonious relationships between humans and animals is told in many of humanity's creation myths, including the American Cheyenne (Preece, 2002):

> *In the beginning the Great Medicine created the earth ... It was always spring; wild fruit and berries grew everywhere ... the Great Medicine put animals, birds, insects, and fish ... Then he created human beings to live with the other creatures ... Every animal ... every bird ... every fish, and every insect could talk to the people and understand them ... people could understand each other and lived in friendship. They went naked and fed on honey and wild fruits; they never went hungry ... During the days they talked with the other animals, for they were all friends.*
>
> *The Great Spirit created three kinds of human beings ... those who had hair all over their bodies ... white men who had hair all over their heads and faces and on their legs ... red men who had very long hair on their heads only. The hairy people were strong and active ... the white people ... in a class with the wolf, for both were the trickiest and most cunning creatures ... The red people were good runners, agile and swift, whom the Great Medicine taught to catch and eat fish.*
>
> *After a while ... the red people prepared to follow the hairy people into the south ... Before they left ... the Great Medicine blessed them and gave them some medicine spirit to awaken their dormant minds ... to possess intelligence and know what to do ... to band together, so they could all work and clothe their naked bodies ... to hew and shape flint ... into arrow- and spear-heads and into cups, pots and axes ...*
>
> *When they went back to that beautiful northern land ... the white-skinned ... people and some of the wild animals were gone. They were no longer able to talk to the other animals, but this time they controlled all other creatures and they taught the panther, the bear, and similar beasts to catch game for them.*

After migrations and ecological disasters during which they 'suffered much and were almost famished ... the Great Medicine ... gave them corn to plant and buffalo for meat ... and the people grew and increased. There were many different bands with different languages.'

The story is not only the memory of what happened to the Cheyenne over many hundreds of years but it 'represents a universal history of mankind' with similarities to many other societies' creation myths. Natural disasters, like the floods of the end of the last ice age, prompted the relationship between nature and other species to alter from the Garden of Eden-like existence. Humans were forced to use nature as a resource, to become carnivores and to value reason over nature or instinct. Preece (2002) states in *Awe for the Tiger, Love for the Lamb* the 'development of such myths indicates that many peoples felt that hunting and eating animals were activities which required a moral justification – or at least a permit from the Great Power'. The myths indicate that while humans, animals and nature once lived as one, ecological changes meant the ideal was now unattainable. However, the initial harmony remained as an ideal; animals to be killed were entitled to respect, especially for their sacrifice in enabling humans to live. The myth resonates with the tensions humans faced, and continue to face, in determining the nature and extent of the relationship between animals and people. Humans share many similarities with animals, but also use them, the tension or conflict between man and animals also one of nature and culture, sentiment and reality, primordiality and civilization. The original condition of mutual friendship between humans and animals separated because of environmental changes necessitating the killing of animals, although they are still worthy of respect.

Humans and animals are socially and ecologically interdependent; changes in one party or the practices it uses and the fate it succumbs to, inevitably have an impact on the other. Some animals depend on non-interference by humans, others dependent on human interventions. In short, animals are important and, either directly or indirectly via our ecology, share common conditions of life with humans. The complexity and diversity of human–animal relationships give rise to different types of relationships meaning the concerns with these relationships are as varied as are the traditions that give voice to those concerns.

Not seeing animals

The Ones Who Walk Away from Omelas, a 1976 essay by Ursula (1929–2018), describes an idyllic society that is dependent on a certain amount of suffering. Accepted by most as necessary for the prosperity of all, the suffering involves a child living in squalor and neglect. Le Guin's skill is in stripping away our lifelong social conditioning to reveal or highlight the issues. In a sophisticated thought experiment, the author invites the reader to imagine the scenes and add his or her own conditions.

The story begins with the inhabitants of the city of Omelas celebrating the Festival of Summer with a procession, dancing, music, singing and children about to race their horses. It is a joyous and serene place without monarchy or slavery, secret police, soldiers or the bomb, but not without the comforts of religion (without clergy), beer, drugs and an orgy if required; it is no goody-good community; the people, mature, intelligent and passionate, lead complex lives. This splendid community, however, depends on a solitary child being kept in a small locked room in a darkened basement or cellar. Feeble-minded, perhaps born defective, perhaps the result of fear, malnutrition and neglect, the naked child sits in its own excrement and suffers. It is thin, its belly is extended, its buttocks and thighs are a mass of festered sores. It used to scream for help and plead, 'Please let me out, I will be good.' No one is allowed to speak or be kind to the child. This is the price of the utter joy and happiness of Omelas that every citizen learns of during adolescence. While knowing of the child and its importance to their and their community's well-being, only some visit it. And only some understand why the child has to be treated so – their happiness, wisdom, and abundant harvests, among other things, depending wholly on it's 'abominable misery.' Idyllic it might be, but not everyone in Omelas was comfortable with the price of serenity. Some boys and girls who visit the child, do not go home afterwards, some adults 'fall silent for a day or two' before leaving home. These people, 'the ones who walk away from Omelas', prefer a place less imaginable and indescribable, perhaps it does not even exist.

Le Guin's allegory or fable can be interpreted as a critique of one of the main parts of our common morality: the costs we justify for our benefits, the willingness to sacrifice something for the greater good, and the rejection of guilt. However, some people cannot face up to or live with the knowledge or the cost and they are the ones who walk away, alone, into the darkness, to an indescribable place, which may not exist. Like the Cheyenne myth, the story highlights that there is a cost to others of our existence, our relationships, one that we accept, respect and do not walk away from. In the human–animal relationship, the benefits are to humans and the costs to animals. Perhaps we cannot live without incurring some sort of debt to animals, perhaps we cannot walk away from them, or perhaps we do not see them anymore. Perhaps it is a lack of moral perception; we don't recognize animals, or at least animal welfare, as a moral problem involving all of us but have left that responsibility to others. There is perhaps no society as deeply alienated or disconnected from other life forms as our own, or as the Haitian proverb tells, 'The goat which has many owners will be left to die in the sun.'

Like the people of Omelas, we accept a certain amount of suffering and, like them, some have forgotten, don't see, can't see, or won't see or tell of our interdependence with animals. Unlike the Cheyenne myth, many have 'forgotten' to respect animals, that they provide our food, labour and many other things. Such a disconnect is partly because, except for pets, the lives and deaths of real animals are no longer part of our lives. Their 'disappearance' means there is no reality check on either our sentiment or our brutality. We remain loyal to the romantic images of the farmyard, but committed to the expediency and efficiency of industrial farming. Eye contact with animals reminds us that we are both like and unlike animals and 'Upon this paradox people built a relationship in which they felt they could both honor and eat animals without looking away' (Pollan, 2002). The loss of everyday eye contact with animals leaves us confused about the terms of the relationship (Berger, 1980; Pollan, 2002). We look away refusing to acknowledge guilt, or become vegetarians and vegans, whereas in the past we have looked and eaten with respect for the animal. What was once a mutual or symbiotic relationship has, for some farm animals, become asymmetric, in that they do not have a good life followed by a quick death. Real animals, especially those in intensive, industrial-like systems, are alien to most of us (Figure 5.4) and, compared with their wild ancestors, some have been likened to living in the equivalent of urban ghettos (Pretty, 2007), even concentration camps (Patterson, 2002).

Terry Pratchett summed it up nicely in *Where's My Cow?* – a book for people of all sizes. 'Why is Young Sam's nursery full of farmyard animals,

Figure 5.4 Individuals in farmyard settings know each other whereas few of us even see, let alone know, the depersonalized industrially farmed animals such as those in barns where it is difficult to even see all the birds (*Photos: Mark Fisher*). A stocking rate of seven birds per square metre led Ruth Harrison (1991) to comment that barns were 'hardly an improvement on battery cages'. (Available in colour in the plate section between pages 178–179.)

anyway? Why are his books full of moo-cows and baa-lambs? He is growing up in a city. He will only see them on a plate! They go *sizzle!*' (Pratchett, 2005). Sadly, farm animals are becoming less visible. A survey of young people (16–23-year-olds) in the UK in 2012 by LEAF (Linking Environment And Farming), a British charity promoting sustainable agriculture, food and farming, revealed that while most knew where food came from, a significant proportion did not:

- 36% did not know bacon comes from pigs;
- 40% could not associate milk with dairy cows and only 41% linked butter to the same animals; and
- while 67% associated eggs with hens, 11% thought they came from wheat or maize.

Furthermore, as adults born in the 1990s, few had visited a farm in more than ten years, if at all. Similarly, in 2011 in New Zealand more than half the general population described themselves as having no contact with the farming community, a further third having only occasional contact (Synovate Ltd, 2011). Isolation surely impairs judgement.

Many people choose not to think about the way their food is produced, the way the animals are farmed, preferring to remain, wilfully, ignorant because they trust farmers and/or have more important things to deal with. Perhaps it is a mechanism to avoid guilt, partly explaining the difference between willingness to pay and actually paying or supporting animal welfare-friendly products. If consumers do trust farmers, and are relatively uninformed of farm practices and animal well-being, supporting farmers to care for animals may be the best strategy for maintaining and improving animal welfare. However, it is a complex issue and it is not known 'whether greater wilful ignorance benefits or harms livestock' (Bell et al., 2017). Willed blindness may reflect factors such as lack of knowledge of details, uncertainty of exactly what animal welfare compromises are involved, misleading claims, feeling disempowered, fearing facing reality, and the uncertainty of how society and individuals transition to new farming systems, or having to change from our present good life (Gjerris, 2015).

If the values we hold reflect, at least in part, our worlds and our experiences, then will people increasingly disconnected from nature, the origins of food and the experiences of animals and farmers, be able to respect and make sound decisions about human–animal relationships (Monbiot, 2012)? Most of us live in urban and suburban environments, our contact with, and knowledge of, farm animals limited to the images on television, the internet or social media. Farm practices are out of our sight, the distance between humans and farm animals reducing caring, responsibility and perhaps the guilt associated with killing. Meat, milk and eggs are the only real exposure we get, but they rarely represent the animal, which is absent, except for idyllic and romantic portrayals on food labels, in children's stories and in our minds. The only animals we regularly interact with are pets, hardly representative of farm animals. However, we still care, perhaps even more so, but without the context borne of knowledge of how our food is produced. Without that worldliness, is it fair or reasonable to expect changes in farming systems to protect animal health and welfare without further impacting on farmers' livelihoods, health and well-being?

In an article entitled 'Young Carnivores Shouldn't Be so Chicken', a UK newspaper columnist bemoaned parent accusations of a teacher traumatizing children by exposing them to wildfowling, the shooting of ducks: 'we have a [supermarket] – people don't need to walk around killing animals to survive any more' (Lewis, 2011). The article went on to claim that you cannot eat an animal without participating in its death, whether it gets 'hooked out of a river, blasted out of the sky or strung up by its ankles and electrified before having its throat slit'. It was claimed that while there are advantages in letting others do the slaughtering, the more you know about an animal's life and death, the more humane your meat eating is likely to be. And letting children grow up not associating meat with an animal's life and death is failing to expose them to their own responsibilities. The article, and the comments it stimulated, encapsulate the psychological dimension of life-dependent resources. Killing animals we depend on, as well as raise, is something most of us have the luxury of not having to face (Figure 5.5).

Killing animals

Animals are a source of food and commerce among other things, but in being so their lives are compromised by loss of freedoms, and by discomfort, suffering and death. There is a tension between our and their interests, between utility and affect, care and cruelty, compassion and violence, loving and killing, understanding and forming close bonds and using and manipulating, that they are like us and unlike us. Humans have experienced and dealt with this tension, for example, with killing animals, in a variety of different ways. From the early hunters to those working in modern slaughterhouses and processing plants, killing animals is difficult; perhaps it is supposed to be (Pretty, 2007).

Humans, mostly, have an aversion to killing, at least during war, and

Figure 5.5 Killing a turkey (Photo: *Reuben R. Sallows, courtesy Archival and Special Collections, University of Guelph Library*).

look to blame or shift responsibility onto something or somebody else. The techniques used include creating a 'distance' to the enemy, making the 'other' despicable, inferior, different or threatening so that it becomes desirable to kill them. The distances can be achieved morally or culturally using ethical, ethnic, racial or religious differences, for example, acting on behalf of others, doing your duty or the right thing and supressing thoughts that killing is wrong. Belief in the inferiority of other classes or castes of people helps, as does using technology that physically distances people, for example, high-power rifles, high-flying bombers or drones operated from the other side of the world (though modern technology also enables the destruction to be more readily seen). The eyes of the killed are the blame of the person killed: 'they shake you more than the streams of blood and the death rattles, even in a great turmoil of dying.' And the language we use also creates distance or raises a veil, for example, 'insurgents are taken out'. These techniques help to suppress feelings of empathy, and closeness or resemblance (Ricard, 2015).

So too with killing animals; it can be difficult (Whiting and Marion, 2011):

> *When Kilchenmann fires, the cow's head jerks upward about a horn's length, the neck arcs back, posture goes haywire and the cow collapses, first on to her horns, then*

her chest, then she finally comes to rest on her right flank. Suddenly she's lying there, so brutally suddenly, without any dignity, smashed down and stretched out on the ground. (Sterchi, 2018)

It can be hard, both physically and emotionally, especially when the animals are young and healthy such as surplus dairy calves. Furthermore, the belief that it has to be done does not make it any easier or less stressful for many people, especially when society and the media assign some culpability and stigma. Killing in modern slaughterhouses may be especially so given the numbers of animals routinely involved (Dillard, 2008). A deer culler reflecting on him and his partner hunting feral goats whose numbers had exceeded the carrying capacity of the environment commented, 'We left the dogs behind and got 175 between us.... It made a considerable difference to our tallies and Bert reckoned it was a terrific day. To me it was plain slaughter, the dull part of hunting' (Crump, 1960). Then there is the slaughter man who recounted to a colleague that there were things he had had to commit in a war-torn country, atrocities few people would understand, yet admitted that there was nothing as mind-troubling as having to continually slaughter lambs. The mass slaughter of animals during foot and mouth epidemics, for example, can result in some people bearing witness to, or responsibility for, the killing, suffering anxiety, guilt, depression and even suicide (e.g. Convery et al., 2005). Similarly, those who interact closely with individual animals, for example, horses and dogs, are reluctant to eat them, those who routinely kill animals for food reluctant to enter into close relationships. Or there is the custom of placing unwanted male goat kids in sanctuaries or pinjrapoles, thereby avoiding anyone having to take responsibility for killing them (see Rodd, 1990).

In historical times, we might have given thanks to the animal, a practice that survives in saying grace before a meal, or reciting the name of God upon each animal before or during the neck cut, part of the Islamic Halal slaughter (see Fuseini et al., 2017). Historically, we may also have misrepresented the animal or imbued it with special qualities, for example, cunning, sly, filthy, personifying it as evil so as to make it easier to kill. For part of our history, the wolf is represented as the devil, big and bad, eating children and old people, yet it is a relative of our close companion the dog. Courage and intellect are used to overcome animals portrayed as powerful and worthy adversaries. We can detach ourselves from animals; in modern times farmers may swap their pet lambs with their neighbours, ensuring they are not personally responsible for their deaths. The animal can become concealed – animals transported, held and killed out of sight of the general population – leading to shock when animals are seen being killed. Finally, we can shift the blame, sacrificing the animals to the gods. In ancient Babylon, in slaughtering animals, the high priest was said to whisper to the victim, 'this deed was done by all the gods; I did not do it' (Serpell, 1986). Or there is convincing the animal that you had just killed that it had met its death by accident (Johnson, 1995). The ancient Greeks blamed the axe or knife for the death of a sacrificial animal or had one person stun the animal and another bleed it, so that neither knew the true killer, sharing or diluting the responsibility (Vialles, 1994). These seemingly increasingly sophisticated psychological mechanisms, rituals, evasions and myths, even acceptance of reality or the need to do so ('people have to eat'), accompanied the shift from traditional hunting ways of life to more intensive systems and exploitive relationships with animals. They enabled humans to distance themselves from or nullify the conflict, guilt or tension between economic self-interest and empathy. We have also distanced ourselves from the people doing the killing; invisibility of the workers once saw them ignored and even reviled. In some cultures, the killing of animals was, historically, the domain of the lowest caste. In Japan, those who washed and disposed of the bodies of the dead and worked in cemeteries were forced to live in segregated communities or hamlets. The *burakumin* or hamlet people would also slaughter and process animals (Milner, 2015). Thomas More's *Utopia* (1516) describes the slaughtering of livestock by slaves because if ordinary people had to do it, it would destroy their 'natural feelings of humanity'. Where once animal sacrifices were solemn events undertaken in public, and animals killed in the middle of towns, slaughter is now 'out of sight and out of mind', the slaughter of livestock having moved from the middle of towns and cities to the outskirts. In some modern professions, for example, companion animal shelter staff and veterinarians, the torment of guilt, grief and frustration associated with having to both care for and then eventually kill the same animals, or kill healthy but unwanted animals, is known as compassion fatigue and is dealt with by strategies such as venting your feelings and altering your emotional attachment, and by having peer support and counselling available (Baran et al., 2009). All these mechanisms acknowledge or recognize the paradox of both identifying with and harming animals.

While animals can be slaughtered humanely (Grandin, 2013), sometimes people act inhumanely. Not surprisingly, human suffering has been endemic to slaughterhouses with workers seeking pleasure in animal suffering (Richards et al., 2013). And yet when they succumb or break and fail to do their best, society generally, as with any form of cruelty, pillories them, and moves to make the installation of CCTV cameras mandatory (Farm Animal Welfare Committee, 2015; Anonymous, 2016). Although there are costs, limitations and concerns, CCTV is considered a useful tool for recording the handling and killing of animals, both to assist with training as well as dissuade irresponsible behaviour, and to assure people that animals are killed humanely, complementing the presence of supervising staff at slaughter plants. Responses to the risk of inhumane slaughter practices also include having rules and regulations, ensuring individuals have the appropriate skills, and undertaking exposés, designed in part to repair reputations and restore confidence in industries. However, those responses might also include greater acknowledgement of the importance of staff, understanding and supporting those having to kill animals, and giving them the status deserving of those who perform a difficult but necessary role. We might ask whether CCTV provides that support, especially when compared with some of the sophisticated psychological mechanisms humans have developed in the past. A willingness to listen to, understand and even help those having to confront the consequences of compassion for animals reflects those age-old rituals that dealt with the tension associated with harming creatures we have empathy with. The alternative is to become indifferent to death, as a means of self-protection. Individuals who must confront the unintended personal consequences of having compassion with animals and having to kill them deserve society's support. There is, rightly, an extraordinary body of work investigating humane slaughter of farm animals (Grandin, 2013). What about the slaughter men? What about the farmers who must cull animals they have milked every day? All those who eat meat give tacit approval to those who care for and eventually kill the animals, however unpleasant those processes are. Should we also care for those people who do our work for us in this way?

A final insight into killing is the effect on ourselves, our community. Reflecting on the banishment of slaughterhouses and killing from modern worlds, Finney (2007) comments:

> *There's something wrong here, and it's a wrong that just keeps getting worse as pressures push wider 'distance' between the meat and its eaters. I understand that people, given a choice, avoid mess and smell and inconvenience. In the world of meat and animals and eating, there's lots of smell and mess. But without the smell and mess, can we understand what's really happening? This anaesthetised life seems so simple. Someone else does the killing, somewhere else, and someone else does the clearing up ...*

The distance between farm animals and most humans, evident in the killing of the former, is long and entrenched. It has increased as the relationship between humans and animals has evolved, the disconnection mirroring the relationship.

Origins of the disconnect

The origin of this disconnection between man and animals is rooted in our past, in the disconnection between humans and nature. As hunter-gatherers, hunters were part of and had a spiritual connection with the land and its animals. The environment had meaning: the trees, the hills, the rivers, the plants and animals, rocks and mountains were alive. Humans were indistinguishable from them and this meant they had to act in harmony with the natural world, respecting the sacredness and power of individual animals. However, while humans should exist in harmony or spiritually entwined with other species, they considered themselves entitled to kill animals. Hunters eased their anxiety with rituals, taboos and prohibitions recognizing that animals had their own lives, mental states and individuality. For example, the Koyukon Indians of Alaska traditionally removed a killed bear's feet first to keep it from wandering (Nelson, 1993). Animal sacrifices, common to religions in antiquity, may have preserved hunters' rituals honouring the animal that gave its life. Rituals forced people to confront and control the discomfort associated with the fact that life depends on the death of others (K. Armstrong, 2009).

However, hunting and rituals were not a good means of providing for a growing human population. Farming began, and with it came new values. While seen to exist in balance with other species, humans were entitled to favour themselves, animals becoming a resource. While God favoured those that helped themselves, animals had to be treated humanely, humans providing for their needs and slaughtering them painlessly. The animal, now part of the human world, was a subordinate and began the path to losing the recognition of its individuality and uniqueness that had been a major part of the hunter's world (Comstock, 2000).

By the time of antiquity, in early Sumerian Assyria and Babylon, Egypt, India and China, humans, now more rationalist and individualist, moved away from using animals to explain their place in the world, and looked to the heavens and science (Wertheim, 1999). Humans, made in the image of God, were no longer animal but had dominion over the earth and its creatures. Preece (1999) explains that the achievements of the arts, sciences, urbanization, prominence of reason over superstition, and technology further contributed to human superiority, the loss of our communitarian natures and the continued rise of autonomous individualism. Along with reason and scepticism, our work ethic, knowledge and education, we have created a vastly different world, superior to other species that 'labour to survive'. This cultural distinctiveness allowed, even encouraged us to forget our biological foundation, that we are animals, reinforced by farming, pastoralism and cultivation being seen to improve or overcome the constraints of nature. Aligned with this, our faith in reason dominated at the expense of other ways of knowing, for example, intuition, history, imagination, senses, feeling, custom, experience, and the older ways of legends and myths, the wisdom of the ages. Reason was seen to free us from our animal origins. Other species lacked reason and were less worthy, altering the foundation of the human–animal relationship. Being human meant acquiring culture and moving away from nature and seeing that same world as a resource (McKibben, 2003). The natural world no longer provides us with the context it once did, we have reshaped it to reflect our habits, appetites and economies. As we have empowered ourselves we have also become more isolated and disconnected as individuals, less likely to see animals and humans as alike, and to identify with, and respect, animals. While all societies possess these elements in varying ways, perhaps western culture does so less than most.

Wendell Berry, a US writer, poet, teacher and farmer, put forward in *The Unsettling of America* that farming is a cultural practice. However, much of modern farming has a different culture, one informed by agribusiness, efficiency and productivity, having shifted from the family- and community-orientated one. It is this shift that is leaving people disconnected or estranged from the land, and from the intimate knowledge necessary for its care. The gradual disconnect between most humans and animals through history is at the heart of the matter. Over time we have lost sight of our relationship with farm animals, our lives missing the details of animals and our relationship to them. How they live, their needs and how they die. Instead, we just see food and our choices of what to eat rarely reflect where it comes from, how animals have lived, how they are farmed and what the real costs are. As the Yoruba proverb states, 'A man of the town knows nothing about farming, or the seasons for planting, yet the yam he buys must always be large' (Ellis, 1894). Or, as Harrison (1971) put it, 'Most people, especially in towns, tend to be ignorant of the processes by which food reaches their table, or if not ignorant they find it more comfortable to forget.'

Modern disconnects

In today's world, the loss of a sense of belonging or connection means there are many circumstances where features differ quite remarkably, as does our understanding. And it is pervasive, we think of the biosphere (nature) and technosphere (man's creation), wealth generated by the former supporting the latter, the latter providing the opportunities and technology for the former.

Many farm animals, as Ruth Harrison pointed out over 50 years ago, have been divorced from the soil, being put into sheds with concrete and other unforgiving surfaces. This separation also extends to their food, access to pasture enabling the animal at least some choice in selecting what to eat. The disconnect is evident between man and nature. Lockwood (2002) tells us, in *Grasshopper Dreaming*: 'The war that is being played out in agriculture is a contest between humans and the natural world to defend our "way of life" – our consumption patterns, our economy, our expectations of comfort, and our demands for convenience.' As farming has developed, disconnects have arisen between it being a 'way of life' and 'a business' and between 'feeding the world' and providing a resource for 'commerce'. While these are played out at farm level, society influences them. But there is also a separation between farming and others in society, be they consumers, animal advocates, major food retailers or whatever. At one level, farmers are usually blamed for poor animal welfare, and they do have the primary responsibility. Should others, the wider food supply chain, processors, retailers and consumers who may care about animals, but do not feel directly responsible for their welfare (Jamieson et al., 2015), also have a more expansive role? They are part of the system determining animal well-being. They are part of the dual subordination of farming noted in history, farmers not only dependent on the climate and ecology of the locale, but also on the constraints imposed by the society in which they live. The dual influences are further represented in a systems diagram (Figure 5.6) suggesting that farming is responding to indirect drivers from the wider community in intensifying production.

I have termed these as proximate and ultimate causes of intensification and the theme of this book is that we need to understand both to ensure the costs and benefits of animal welfare are distributed equitably and sustainably. If farming is both driven by and responding to societal factors, then at least we should acknowledge their influence and, if appropriate, explore the possibility of improving animal welfare by making changes to the indirect drivers, those coming from beyond farming.

Is the disconnect exacerbated by not 'showing' reality, for example, the killing of animals. How can farmers incorporate society's values if society is not aware of the realities of their roles and responsibilities? Would animal welfare be more sustainable if we did show or tell the reality? Is failing to do so akin to evolution's fatal flaw, accepting or gaining short-term benefit, even if the gains are detrimental in the long term? The tyranny of the minority – social media representing the abnormal as normal – suggests greater exposure to the realities of farming may have some risks, at least initially.

Otherwise, by not ensuring people are aware of the realities of farming,

People are people through animals • 197

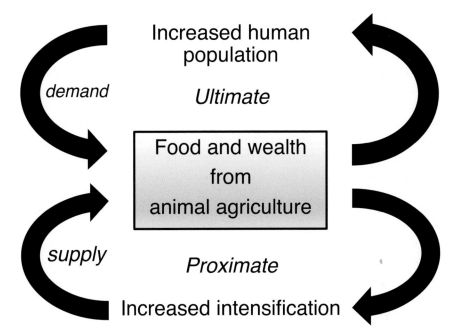

Figure 5.6 A systems representation (Senge, 1990) of agriculture changes represented in a cycle, one enabling and driving the other. It is suggested that concern for animal welfare is a symptomatic solution and that to address animal welfare in a more equitable and sustainable manner society should also consider more fundamental solutions. The two cycles, the higher one human culture, the lower one biology, make up agriculture. The two cycles represent the interaction and the disconnect, the need to balance or align, not just the farm (lower) but the processing, distribution and consumption (upper) parts of both commerce and the consumption of products.

are we disconnecting society from the natural world? This is important since one view of ethics is that it is the systematic reflection of moral issues in the public sphere (Lassen et al., 2006a). It is clearly difficult to reflect on these issues if the audience to be engaged with is unaware of them. On the other hand, does not showing them provide a mechanism for people to deal with the tension of killing animals, especially very young ones, which humans may have a strong biological affinity with (Mellor et al., 2010)? But if people cannot cope, are they happy to entrust reality to someone else? If that is the case, then trust, acting for the sake of those who entrust, as well as for personal and institutional gain, is the issue that must be addressed. We have become removed from the realities of farming, so far removed that any reminder is disturbing (Monbiot, 2016). The unfamiliar attracts condemnation, shock and horror. Perhaps if we engaged more, understood better, the natural world, and our place in it, we could develop a more sustainable vision of animals and their welfare. As Harrison (1964) noted, 'we must cease to pander to an unenlightened public.' If practices are both reasonable and necessary, and justified, then the public need to be reassured.

These dilemmas are made more complex by the realization that animal farming may not mean the same thing to different people. Many do not express their concerns as citizens through their behaviour as consumers. To some, farming is seen as more than just food production, people wanting farms to be traditional, modern and natural, too. The challenge of expecting to resolve and maintain the desired balance between different worlds or perspectives is at the heart of sustainability: economic, ecological and sociocultural (Boogaard et al., 2011). Others are aware of the difficulties and are willing to compromise but need to see the combining of the best of the three worlds in order to be sustainable from a sociocultural viewpoint. Thus, the best is socially and culturally constructed, and dynamic. People may accept modernity at the expense of the natural, but some things, for example, increasing stocking densities, decreasing farmer–animal contact, decreasing family farms or putting economic interests above animals, may be unacceptable.

Dealing with these dilemmas requires society to engage in sophisticated conversations, unlike 'the most shrill and dramatic articulations' that tend to demand attention, or those seeking 'to privilege the transient urges of the mob over and above social order' (Norman, 2013; Rollin, 1996). In effect, we must adapt our cultural expectations to suit and guide farming. Humans do change their cultures, often in ways that impact on animal welfare. Three examples, affecting rabbits, sheep and beef cattle, demonstrate the impacts of culture on farming, and thus the opportunity for culture to address issues like animal welfare, which has traditionally been seen as the domain of the farmer.

Changing practices

Just as sheep are entwined with shepherds, so too are rabbits linked, both in farming and in the wild, with man, whether as farmers, sportsmen, consumers or politicians. The rabbit has been introduced, protected, controlled and killed such that 'the history of the rabbit is the history of man, and his impact on plants and animals, farm crops and livestock' (Sheail, 1971). Rabbits are valued as wildlife, as pets, farm and research animals, and at the same time almost despised as a nuisance, pest or competitor for humans' economic and conservation interests. While humans and rabbits coexist, the former makes a wide variety of uses of the latter, but it is the benefits of the latter that have changed through history that make the rabbit a good example of human culture impacting on animals (Bailey, 1988; Druett, 1983; Lebas et al., 1997; Sheail, 1971; Thirsk, 1997; Veale, 1957).

Once a native of the Iberian Peninsula and northern Africa, the rabbit is among the most successful animals. Beginning in antiquity, the Phoenicians, Romans and Greeks spread it, and more recently, with European colonial expansion, it has very successfully adapted to countries such as Australia and New Zealand. Through much of the rabbit's history, it has been a rare and prized commodity highly valued for its meat and fur, a 'fashionable delicacy of the rich'. Rabbits were kept in leporaria or stone-walled pens and parks, the origin of the warrens or game parks of the Middle Ages. They found their way to England from France by the 13th century where they became established, the earliest warrens artificially created by warreners constructing 'rabbit-buries' or artificial burrows or 'pillow-mounds' including ditches to prevent drowning. In feudal systems, the right to hunt and kill any beast or game was a special privilege and keeping and killing rabbits were 'gentlemen's enterprises', illegal for anyone else. For many centuries the rabbit lived protected within those warrens, sometimes provided with food during harsh winters and droughts. It was not until the 18th century that the rabbit colonized a much wider area because its predators were reduced by landlords protecting other small game (e.g. pheasants and hares) and starvation was minimized when farmers replaced winter fallow with crops for sheep and cattle. The wild rabbit found an ecological niche in farming. While rabbit was relatively scarce, it became a favourite dish on the tables of the great and wealthy, costing four to five times as much as chicken. Income from rabbits also contributed significantly to manor incomes. One warren in the UK produced an average of 29,000 rabbits a year during 1855–1862. However, as numbers increased, rabbit became available at markets and prices fell from 3 shillings and 6 pence over the 13th century to 1 shilling and 8 pence during the second half of the 15th century. While still a relatively expensive commodity, prices could not offset the continuing rise in wages and transport costs and the industry would have suffered from declining profit margins. The management of warrens was a skilled undertaking and good warreners were valued, especially given the threat from predators and poachers, and paid well. These threats, and the determination of landlords to protect their assets saw the construction of warren watchtowers (one with three-feet-thick walls and slit windows) or lodges, landlords managing their warrens with great care. In one 15th-century instance, three priests were caught poaching with their own specially reared greyhounds and fined a substantial sum of money. Regular poachers reared their own dogs and ferrets and made their own nets, and organized themselves into gangs with a range of poaching accessories (soldier's tunics, steel helmets, bows and arrows). By the 16th century, the rabbit had begun to be domesticated although wild rabbits were kept for the symbolic value of hunting, and the power and prestige of nobles (Monnerot et al., 1994). As poachers made rabbit available to the poor, and rabbits spread, so becoming more accessible to peasants, rabbit descended the social scale. Although legally excluded, many peasants living near warrens would have secured a supply for consumption or distribution through the black market. New markets developed, and rabbit replaced Russian squirrel as the basic fur of northwest Europe. Many warrens were dismantled to make way for more profitable grain growing in the 18th century. Changing land use, legislation allowing farmers to catch rabbits on their own land, and the importation of rabbits, meat and skins from other countries brought an end to warrening. By the 19th century, although the rabbit damage to pasture and crops was acknowledged, it seemed it was outweighed by the value of the fur and meat (the 1960 edition of the iconic *Mrs Beeton's Cookery and Household Management*, arguably 'the world's greatest cookery book', had some 40 rabbit recipes).

It was from this background that rabbits were introduced into New Zealand in the first half of the 19th century. Once established, and with a lack of natural enemies, they became enormously successful, infesting large tracts of the country. Occupying the same ecological niche as sheep and cattle, they ate 'the land bare' and bankrupted many grazing operations. Rabbits became big business and full-time rabbiting (Figure 5.7) became so profitable, rabbiters paid property owners for the privilege of cleaning out their properties (McLean, 1966).

The advent of refrigerated shipping in the late 19th century further made rabbit meat a profitable business. Eventually a panoply of measures, including hunting, shooting, poisoning, the introduction of predators and diseases, trapping, the establishment of rabbit destruction authorities, prizes, and taxes to make surreptitious farming of rabbits unprofitable, resulted in a satisfactory decrease in rabbit numbers. The aerial application of poisons killed large numbers and led to public outcry, both for and against the rabbits, leading to the comment, 'We don't poison rabbits with phosphorous because we are "cruel". We'd rather kill them painlessly if this was possible: no sane man gives a rabbit phosphorus for fun. But if we must be cruel to the spoiler, to prevent him spoiling our heritage – well then we must be cruel. You cannot have both unchecked rabbits and productive farm lands' (McLean, 1966). However, this was not to last; the events leading to the illegal introduction of rabbit calicivirus, as outlined in Chapter 1, show that rabbits and people still have a complex and changing relationship. And the future may be no different; likened to a 'biological refrigerator', a rabbit is able to be consumed at a single sitting without the

Figure 5.7 A rabbiter skinning rabbits for drying in Central Otago, New Zealand, and an advert from 1940 for selling skins (*Lakes District Museum*).

need for processing and storage, so is considered a suitable animal species for farming in parts of the developing world (Boland et al., 2013).

Shepherding during lambing is another example of animal husbandry being shaped by human culture. The period during which an animal gives birth is notable for the complex interactions of maternal and foetal/newborn physiology and behaviour. These include the ewe, normally a social animal, seeking isolation that may have once reduced the risk of predation, and aided in the exclusive bonding of dam and offspring. Should sheep be shepherded so that difficult births can be assisted, vaginal prolapses treated, moribund animals destroyed, displaced ewes and lambs reunited, and orphans fostered, artificially reared or euthanased? Or, conversely, should shepherding be kept to a minimum, so that animals can give birth in relative isolation, without human interference contributing to difficult births and lamb mortality? Those responsible for sheep during lambing face this dilemma, and how they do so depends on many things such as what resources and labour are available, the environment and the genotype of the sheep. However, a major influence is also the cultural expectation to shepherd, with one modern farmer lamenting that, 'it appeared as if everyone had been brought up with the "good shepherding" tradition and was prepared to work all hours ...' One approach was to select sheep with a high survival rate and low lamb mortality, that require less shepherding at lambing than others. Many factors have been identified as contributing to the 10–14% improved lamb survival including larger pelvic openings, a lower incidence of difficult births, and better maternal behaviour (Dwyer and Lawrence, 2005; Knight et al., 1988).

Shepherding was a cultural legacy and not the long-term solution to some of the problems associated with extensive pastoral farming. The pragmatic approach was to breed stock to suit their environments, as well as to adapt stock handling skills and farm management practices. Several themes were apparent in this approach. Animals were bred to survive or suit local environments or conditions, particularly steep hill country. This involved extensive culling of undesirable animals, regardless of how well they might perform in traits other than the ability to survive and to produce live lambs. Both artificial and natural selection were used, the latter enabling the important traits to be identified and subsequently incorporated into artificial selection programmes. The impracticality of supervising lambing in difficult terrain and the cost of skilled farm labour necessitated the practice. Finally, it was acknowledged that disturbance at lambing created problems and, most importantly, the approach reduced some of the problems traditionally associated with lambing.

Strategies associated with lambing vary, from the intensive indoor to extensive outdoor systems, and from virtually continuous supervision to minimal or no shepherding. One of the major differences in lambing management systems is in the value placed on human supervision. Few studies have compared the merits of different shepherding systems, but it has long been known that the ewe may attempt to seek isolation around the time of birth, and that disturbing or removing her from the birth site, the focal point during lambing, can significantly increase lamb deaths (Putu et al., 1988). In more intensive systems, the benefits of shepherding are paramount while in some extensive systems the costs of shepherding are fully acknowledged. A central feature of the latter system is the view that human interference can cause harm by inhibiting or delaying parturition, thus increasing the risk of dystocia as well as compromising ewe–lamb bonding. This may be especially so in extensive farming situations where the animals are not well habituated to regular contact with humans.

There is a rich cultural legacy of sheep and shepherds, something reflected in the biblical statement, 'the good shepherd giveth his life for the sheep' (John 10:11). Historically, the relationship between humans and sheep was probably based on small numbers of animals and may have meant that intensive husbandry was practical if not imperative. Shepherds may have had to watch their flocks by night, guarding them against predators and features of this relationship remain in some parts of the world. Modern sheep farming is, however, largely characterized by large flocks and a lack of predators, suggesting that part of the demand for intensive shepherding is a cultural vestige from historical farming practices. Have we created the modern sheep, and sheep farming system, but retained the historical shepherd?

The vast and endless pampa plains of South America have been home to the gaucho, the legendary horsemen, and their extensive cattle herds. Raising cattle was part of Argentina's cultural and economic history, local beef consumption per person and exports some of the highest in the world. The land and its native and cultivated pastures were well suited to grass-fed production, a distinctive feature of Argentina. Few cattle, until recently, were supplemented or finished with grain. But the days of the gaucho and extensive cattle farming on the vast pastures are waning.

Although several factors have contributed to the changes over the last few decades such as increasing soybean prices, severe droughts in marginal areas, more modern varied diets, lack of investment in the industry and currency devaluation, the Argentine government, in an effort to lower rising prices of beef, banned exports for six months and then imposed a 15% export tax on fresh product. While domestic beef prices dropped, ranchers and farmers reduced their herds, finished cattle in feedlots and replaced their pastures with more profitable soybean, corn and wheat. Where cattle were once encouraged onto feedlots by government subsidies, they are now maintained by increased beef returns. Government policies and soaring beef prices have led to the once unthinkable: the importation of beef. Beef is embedded in the national psyche, its quality lauded, and there is debate amongst chefs over the taste of the beef: the more intense flavour and greater texture of meat from grass-fed animals or the tenderer, juicier but less flavoursome or different-flavoured feedlot-produced beef. While 'old gaucho grieve for the past', Argentina is following more efficient systems used in other beef-exporting countries, feedlot beef not only set to stay but to increase in future (Forero, 2009; Gonzalez, 2016; Misculin, 2008; Queck, 2013; Rizzi et al., 2016).

These examples of changing relationships between humans and animals reflect a changing human perspective; changing cultural expectations of the value of the rabbit, the abilities and needs of both sheep and shepherds, and the move from pasture to feedlots in Argentina, all had consequences for animal welfare. The complexity of the relationship between mankind and animals is seen not just as a concern for animal welfare, but also as a stimulus for the continuing co-evolution of mankind and animals. It is suggested that there will continue to be a place for different relationships to form, since each is guided by different factors. Demands of animals on us and our perceptions of their needs vary with the complexity of our morality. Inevitably, different people will treat animals differently. Consider three pictures of what farming could be like, all of which produce food and provide economic livelihoods, yet evoke different responses in different people for different reasons. The first is the romantic, but largely historical image of farming, where chickens roam the farmyard and cattle and sheep graze green meadows. The second is conventional farming and includes traditional extensive pastoral farming as well as the more intensive feedlot and factory systems common in much of the western world. Cows might be grazing extensively on pasture and be milked twice a day; or continually housed, milked three or more times a day by a robot, and fed concentrates to match the quantity of milk produced, perhaps genetically modified so as to not suffer from production diseases. Similarly, poultry systems vary markedly. We find the image of free-range hens aesthetically pleasing, and some people vouch for the quality of their eggs and meat. And finally, there is the third image, of genetically modified bacteria producing 'cow's milk' in an industrial vat. Or other futuristic images of farming including fast-growing trees that are pulped and digested into simple sugars for genetically engineered bacteria to convert into more recognizable food (Anderson, 1990); algae in slime pits converting solar energy into protein that genetically engineered bacteria convert into food (Rogoff and Rawlins, 1987); or cell-grown meat and milk (Pacelle, 2016). Consider 'Chicken Little': an ever-living slab of chicken flesh, from which any number of slices can be cut, in essence an edible tumour. However aesthetically displeasing, it provides inexpensive meat with minimal cost to the welfare of birds. Science fiction (Clark, 1998) has thrown up many other wonderful examples including Douglas Adams' 'Dish of the Day', a creature ludicrously eager to be killed and eaten.

All these systems produce food and provide opportunities for commerce, yet mean different things for humans and animals. While the futuristic scenarios might produce safe food in environmentally sustainable ways, as well as present business and career opportunities for biotechnologists and others, what impacts will human culture have on their success, and what impact will they have on human culture?

All these systems produce milk, meat and eggs, yet which is best for the cow, the chicken, the farmer, the biotechnology company, the consumer or society in general? Do we want to be the sort of people that give up the pastoral or free-range ideal, an animal living reasonably contented, or an industrial one? Can we afford not to? What sort of animals will we want to live with? What sort of innovations do we want? What sort of food do we want to produce and consume? Does it matter? As a society, we need to start exploring questions like these for they will determine how humankind evolves. Many cultural and biological factors determine how we relate to animals including their needs and our innate or culturally ingrained beliefs and prejudices that determine how those animals are treated (Johnson, 1995). One such factor known as biophilia, or a love of life, is a natural disposition to be interested in and attracted to other living creatures (Kellert and Wilson, 1993). Animals, plants and nature are important to humanity in some way beyond their instrumental value as food or sport, for example. We take delight in the real presence of an animal, a mountain range or a garden park. A variation on this theme is 'the necessity of wildness', something inherent in our dedication to preserving National Parks in some pristine mankind-free state. Thus the method of tending to or husbanding food animals might have some intrinsic value to humans, independent of the food and incomes they produce.

The affluent contract and the burden of animal welfare

Farmyard hens usually range freely, often accompanied by roosters, and are able to dust bathe, access water and varied food, and even roost in trees (Figure 5.8). The romantic ideals of the farmyard or free-ranging hen contrast markedly with the reality of caged layer hens in industrial-like systems.

Yet the former is a life that does not provide humans with sufficient benefit, in terms of plentiful and inexpensive eggs efficiently, for it to be a modern and widely accepted source of food and commercial livelihoods. These hens lay fewer and more expensive eggs and, despite being cognizant of welfare, most of us value plentiful and inexpensive eggs over the welfare of the birds. Or more accurately, although we care for bird welfare, we do not care enough to value and pay for systems that give the birds a good quality of life. As already noted, while a small number of

Figure 5.8 Farmyard hens and roosters living (what we might imagine to be) an idyllic life (*Photos: Mark Fisher*).

people may be willing to pay large amounts to improve animal welfare, moving hens and pigs from confined to free-range systems, the majority would pay little or nothing (Lusk and Norwood, 2012). Although we may believe animal welfare is important, or is becoming increasingly important, the authors suggested that 'humans generally place a low value on animal welfare'. While it is acknowledged that the insights provided by studies assessing willingness to pay are contentious (McInerney, 2004), it is tempting to conclude that few value animal welfare, at least from an economic perspective. As many have alluded to, consumer concern for animal welfare is, arguably, not matched by actions, the burden of caring for animals residing with those in charge of them, despite beliefs in wider roles or responsibilities. For example, 'We, the consumers, cannot lay blame for poor animal welfare at the feet of farmers. The problem and the solution are in our own hands' (Webster, 2005a) and, 'It is important to remember that animal welfare, whether pastoral or other animals is a total society responsibility' (Kilgour, 1985). This is not surprising. For many people in the modern western world, interactions with animals tend to be limited to emotional ties of affection with their pets, arguably more akin to members of the family. In the absence of understanding and empathy associated with reasonable and necessary use of farm animals, some representations, such as the farmyard hen, have attained a detached level of significance. We remain loyal to the farmyard hens but committed to the products of industrial cages (Strange, 1988). Commonly held beliefs, rather than personal experiences, potentially determine the character of society's expectations of animals. It is interesting to contemplate what imagery we might draw on in future: our pets, or farm animals of largely bygone eras retained in our myths and beliefs (Lawrence, 1993)? Or, like the people of Omelas, do we accept some suffering?

This is the nature of our modern 'affluent' relationship with animals. In contrast to the 'ancient or domestic contract' – we look after the animals and the animals look after us – an 'affluent contract' might be imagined: we value the products of animals and they look after us. This contract is founded on plentiful animal products, produced as efficiently or as cheaply as possible, providing a source of commerce for many people in the food supply chain, as well as food and other commodities. Aesop's fable of the *Goose and the Golden Egg* tells of a countryman, hoping to become even richer, killing a goose to find the source of the golden egg it laid each day. The fable is the source of 'killing the goose that lays the golden eggs', a phrase warning of the perils of greed. To paraphrase Aesop, the 'golden eggs' have become more important than the 'goose', in part because we do not see or know the goose anymore and do not see or appreciate our dependence on it, and in part because we do not allow it to flourish. Table 5.3 highlights that the benefits and costs of the relationship between animals and humans are weighted in favour of human flourishing, and in the case of farm animals raised in industrial-like systems, flourishing is significantly reduced.

In a utopian relationship, both humans and animals would flourish but we cannot live without having an impact on animals, and utopia, like 'the wolf lying down with the lamb, the leopard with the goat' (Isaiah 11:6) is, by definition, unobtainable. However, it does not mean that we should not strive to improve animal flourishing. Rather, it is the costs of those improvements, and who bears those costs, which are at the heart of the future of the relationship between humans and farm animals.

Much of the western world's affluence is borne of, at least initially, the domestication of animals and plants and their subsequent intensification. Food production is now so efficient that relatively few people are directly involved, though many more are involved in processing and distributing it, and in supporting its production. We spend as little as 10–15% of our income on food. However, we also care about animals; affluence has enabled us to question the methods of raising farm animals, and to develop aspirations of more equitable, even revolutionary, relationships, and it is a source of tension for some people. This is partly because some animals

Table 5.3 A summary of the characteristics of humans, in most western world countries, and farm animal flourishing (adapted from Nussbaum, 2004; Tulloch, 2011). While many align with the needs of animals (or the Five Freedoms), others reflect the common view that it is acceptable to use animals if that use is humane ("has pleasure in life while it lives and then is humanely slaughtered; Harrison, 1971). For example, while humans are generally able to live a normal-length life, most farm animals are killed well before their normal life expectancy. (Y = Yes, N = No).

Characteristics of a good life	Humans	Wild animals	Farm animals Extensive	Industrial
To be able to live a normal-length life	Y	Y but varied[1]	N	N
To have good health and to be adequately nourished and sheltered	Y	Y but varied[2]	Y but varied[2]	Y
To not be disfigured	Y	Y	N	N
Having freedom of movement, security from violence, and opportunities for reproduction	Y	Y but varied[3]	Y	N
Being able to use the senses to imagine, think and reason and have pleasurable and avoid harmful experiences	Y	Y	Y	N
Can form emotional attachments with others	Y	Y	Y but limited[4]	Y but limited[4]
Live with and show concern for others and be treated with dignity	Y	Y	Y	N
Live with and show concern for other species and ecosystems	Y	Y	Y	N
Can play	Y	Y	Y	Y but limited[5]
Have control over one's environment	Y	Y	Y	N

[1] Wild animals are normally subject to predation. [2] Varied health and shelter in different populations of wild and extensively farmed animals. [3] Wild animals are often hunted, or their habitats defined by humans. [4] Farm animals are usually weaned at an early age and kept in age- and sex-specific groups. [5] Play constrained in barren and confined environments.

are not flourishing, partly because we do not know them, but depend on ideals, beliefs and myths to inform our expectations, some of which may be outdated and in need of scrutiny. It is also partly because expectations of the relationship are revolutionary in that they do not reflect the common view.

The 'affluent contract' is founded on greater 'use' of animals than the more mutual relationship of the 'ancient contract'. Furthermore, the burden of providing welfare is being left to fewer and fewer people operating in a system that does not, and arguably cannot sustainably, address animal welfare (or farm incomes, and the environment). This burden is borne of not only caring for animals but caring for them while at the same time making them produce more and faster with fewer resources.

Changing the nature of the relationship is not just a case of educating people, or exposing them to farming and farm animals. Reflecting the diverse understanding of animal welfare, exposing people to farming is also unlikely to resolve people's concerns (Ventura et al., 2016). Similarly, educating people about contentious issues may only serve to polarize them. Although disconnected, some at least refuse to blame the farmer, acknowledging the mostly economic and structural factors, including themselves (consumers), in sustaining poor systems through purchasing cheap food. While consumer demand for cheap and plentiful meat, milk and eggs may have helped develop intensive farming, farmers, industries and governments chose to meet that demand with such systems. Historically, there has been 'a lack of questioning prophets' (Midgley, 1991) until Ruth Harrison's devastating critique of factory farming brought it to the attention of the wider community. What makes a life go well, what enables animals to flourish and what is a fair relationship between animals and humans? Having in place farming systems that result in the acceptable welfare of animals is a task that is best shared by all: farmers, the food supply chain, scientists, governments, consumers and the public. While farmers have the primary responsibility, it should not be theirs exclusively.

We might assume, as many have, that the use of animals is acceptable if it is humane, and that vegetarianism is not a realistic objective. Furthermore, we cannot turn the clock back, for example, lowering inputs and productivity. Animals have changed, as have farmers' knowledge and the systems and social environment they inhabit. A more sophisticated response is required. Enforcing changes to improve welfare without acknowledging the forces driving intensification may do more harm to animals, farmers and society. Perhaps we should design a system that rewards, or at least fairly compensates, farmers and animals for working towards the ideals society accepts. In essence, we have to confront moral problems in the context of preserving relationships, or developing the relationships society wants. This may require acknowledging the impact that the influences of urban populations place on farming and farm animal welfare. What are the values of the affluent contract (Te Velde et al., 2002)?

The 'growing importance of animal welfare as an arena of political, market and social intervention demands a greater understanding of the shifting influences of the market, of legislation and of farm animal welfare as a public good' (Farm Animal Welfare Committee, 2014). The system itself must flourish, not just the animals, farmers, processors, retailers, consumers and others. The inequity in the relationship is between the two cycles in Figure 5.6; we address animal welfare at the lower or farm level because that is where the animal is but overlook factors impinging on the drivers of animal welfare coming from the upper level. Furthermore, when we critique, examine or question the lower level it is often from a limited perspective. Are we too scared to understand the complex picture? Is this reluctance a psychological mechanism to deal with caring and killing? Is it naturally easier to lay blame or tell others what to do? The animal welfare industry may have the right sentiments, but will it impact on farming by imposing greater costs when it should arguably be understanding and sharing them, enabling or facilitating good animal husbandry even if it comes at our expense? The next chapter explores the opportunities for restoring the culture to agriculture, the possibility that changes in society, where most of the people are concerned with animal welfare, might also be able to enhance animal welfare since animal welfare standards are a reflection of how society values animals.

Further reading

The Unsettling of America: Culture and Agriculture, 3rd edn. **Wendell Berry**. Sierra Club, San Francisco, CA, 1996.

Sacred Cows and Hot Potatoes. Agrarian Myths in Agricultural Policy. **William Browne, Jerry Skees, Louis Swanson, Paul Thompson and Laurian Unnevehr**. Westview, Boulder, CO, 1992.

Barriers to the fair treatment of non-human life. **Andrew Johnson**. In: D.E. Cooper and J.A. Palmer (eds) Just Environments. Intergenerational, International and Interspecies Issues. Routledge, London, UK, 1995, pp. 165–179.

Food Wars. The Global Battle for Mouths, Minds and Markets. **Tim Lang and Michael Heasman**. Earthscan, London, UK, 2004.

The End of Nature. Humanity, Climate Change and the Natural World. **Bill McKibben**. Bloomsbury, London, UK, 2003.

Animals Strike Curious Poses. Essays. **Elena Passarello**. Jonathan Cape, London, UK, 2017.

The Earth Only Endures. On Reconnecting with Nature and Our Place in It. **Jules Pretty**. Earthscan, London, UK, 2007.

Animal to Edible. **Noëlie Vialles**. Cambridge University Press, New York, NY, 1994.

CHAPTER SIX

Thinking like a mountain

The fierce green fire in the eyes of a dying old wolf reflected something known only to her and the mountain. The hunter sensed that his own understanding, fewer wolves meant more deer, did not have the wisdom of the wolf and the mountain. In the wolf's eyes, Aldo Leopold (1887–1948) saw the need to think differently, that human actions and values should acknowledge the interdependence of animals, the environment and humans, good behaviour being that which sustained the integrity, stability and beauty of the land. Leopold was employed to manage the forest, principally for recreational deer hunters, so he set about killing the deer's predators, the wolf and the mountain lion. However, the deer population, no longer held in check by predators (including human hunters), exploded resulting in severe damage to the forest and the land. As his interests grew to include conservation and ecology, Leopold concluded that the forest feared the deer more than the wolf. As 'only the mountain has lived long enough to listen objectively to the howl of a wolf' we should 'think like a mountain' (Leopold, 1949). Like Leopold's wish to protect the deer for humans, our desire to protect the well-being of farm animals is understandable. However, traditional approaches to animal welfare are difficult; it is more than 50 years since *Animal Machines* was published and modern societies are still struggling with intensive or factory farming. Perhaps society needs to change, to think more like the mountain than the deer.

Another American, cancer researcher Van Rensselaer Potter (Figure 6.1), extended Leopold's environmental ethic from the importance of the maintenance of a healthy ecosystem to the recognition that the behaviour of humans must be coherent with ecological realities. In understanding the need for biological wisdom, Potter coined the term 'bioethics', urging the reconnection of science and the humanities. He claimed that ethical values cannot be separated from biological facts (Potter, 1970) if we are to overcome our biological predilection for short-term gains that may be disadvantageous or fatal in the long term. Unless able to 'forestall the onrush

Figure 6.1 Van Rensselaer Potter (1911–2001) at the 'Stuga', his writing hut and retreat in the woods. Potter, who first coined the term bioethics, described the need for a new wisdom based on knowledge, human values or ethics and biology including medicine, the environment and agriculture (*Photo courtesy Lisa Potter Bonvicini*).

of unthinking and self-centred economic activity', Potter believed or was concerned that our species would be decimated by the failure to recognize the environmental conditions necessary for our continued existence (Potter, 1990). Right and wrong were ultimately defined by our survival in a hospitable or healthy environment and since biology is variable and can be regionally specific, then human values will vary, depending on the biology on which they are based. Thus, no matter how our ideas of right and wrong are informed by universal patterns of thoughts and beliefs, different ecological systems may mean diverse values and consequently different treatment of animals: a sort of moral ecology (Donnelley, 1995).

The ecological literacy Leopold and Potter advocated is to understand and live according to the limits of biology, our social world and the physical environment. As Auden (1907–1973) put it, 'A culture is no better than its woods', society part of a socio-ecological system, its health reflecting its biology. Many peoples have recognized and continue to recognize the connection, but 'Only in the last moment of human history has the delusion arisen that people can flourish apart from the rest of the living world' (Wilson, 2001). History, and an understanding of the changing relationships between animals and people, suggest that human culture has grown too big and become disconnected from biology. Farm animal welfare is then, as well as the state of the animal, what it feels or experiences, a reflection of human culture.

Culture and agriculture

The perils of human culture are illustrated in Olaf Stapledon's (1930) masterful science-fiction account of the rise and fall of distinct races or species of humans. *Last and First Men* tells of mankind's growing culture, 'the spiritual development of the world-community', increasing the earth's gravitational pull on the moon. Drawn closer and closer to the earth and eventually disintegrating, the heat generated from lunar fragments crashing to the earth would have made life impossible, necessitating relocating to another planet. Later men would have to 'preserve the solar system from confusion' from their 'greatly developed mental and physical activities' by continuously expending energy to counter the gravitational pull.

Like the attraction between human culture and the moon, it is suggested that human culture is exerting a pervasive influence on farming, such that animals, farmers and the environment are, in some situations, being pulled beyond their limits. For example, making animals more productive or increasing stocking rates, to ensure or maximize income. While we may believe that 'biology keeps culture on a leash' (Cochran and Harpending, 2009), there is a lag phase meaning culture is no longer constrained. The domestication of animals and the growth of the urban human population have reduced, or hidden, our dependence on biology. Human culture has grown, concealing, for many people and for many parts of the social-ecological system, human dependence on biology, agriculture and farm animals. Leopold believed, 'We can be ethical only in relation to something we can see, feel, understand, love, or otherwise have faith in' but in our modern world, many do not have those connections with, or sense of belonging to, animals. Others have also argued that contact and experience lead to a better understanding, empathy and emotional attachment (see Cornish et al., 2016). And this is evident in society's responses to poor animal welfare. Typically, increasingly higher standards are expected and enforced, standards that are underpinned by knowledge and technology but that inevitably come with a cost. All the while, returns from animal products generally and steadily fall (see Chapter 3). In addition, expectations that animal welfare is assured, monitored or verified place additional costs on farming. Currently, such costs can only be covered by intensifying production and with it higher demands of stockmanship and increased risks of failure. To return to Leopold, the deer is not only being protected but being made more productive.

If, as the evidence and stories raised here suggest, some instances of poor animal welfare are the symptom of a system based on intensification, then attempts to address it in the way that we currently do are unsustainable. To paraphrase Edmund Burke's (1729–1797) comment, 'We whip the child until it cries, and then we whip it for crying' (cited by Saul, 2005), we push farm animals, farmers and others until they fail, and then punish them when they fail. Similarly, as Aleksandr Herzen (1812–1870) said in speaking to a group about how to overthrow the Czar, 'We think we are the doctors. We are the disease.' Finally, Leo Tolstoy (1828–1910) captured it in *What Then Must We Do?*, a work about social conditions in Russia: 'I sit on a man's back, choking him and making him carry me, and yet assure myself and others that I am very sorry for him and wish to ease his lot by all possible means – except by getting off his back.' Furthermore, the more we address the acute or proximate problems, the more we contribute to the chronic or ultimate conditions that contribute to them (Covey, 1989). Animal welfare, then, is a symptom, or a reflection, of the well-being of humans, human welfare paid for by animal compromises.

So we come to that stage of a book when the author offers wise and profound solutions since any criticism must naturally be accompanied by suggestions for a better way, one that is both practical and achievable. There have been numerous such suggestions and visions of what farming might and should look like in future, many of them driven by a concern for animal welfare. They include agro-ecology, community-shared agriculture, small-farming, farmers' markets, precision agriculture, traditional farming, enlightened agriculture, sustainable intensification, family farming, technology-enabled agriculture, silvo-pastoralism, high-yield and high-technology agriculture, multifunctional farming, redesigned agriculture, organics, farming insects and making greater use of other novel sources of protein including plants and algae, sustainable agriculture, ethical accounting, cutting out the middleman, enlightened consumers supporting innovative practices, developing niche and novel products, plant-based cultured or lab-grown meat and milk, smart farming, reducing meat consumption, dietary changes, increases in farming efficiency and so forth. These and other initiatives, all valuable in themselves, do not address the ultimate source of the problem, human culture outgrowing biology: with the expectations that food is inexpensive, animals are cared for and that, increasingly, their welfare is assured if not monitored or verified. The possibilities are very extensive. Nor is a single solution expected or likely to ensure or realize a more sustainable social-ecological system. In one respect, this book fails, for I do not have a neat solution like the examples in this paragraph. Instead, I suggest that we explore things differently, armed with an understanding of the important interconnection between animals and people, and an acknowledgement that animal welfare is a complex or wicked problem requiring something different to the exposés, regulations, education or promises of premiums, important as they are. And it will require the involvement of more than just farmers and others telling farmers what they should do, how they should manage their resources. Ultimately, it may require behaviour change by all – consumers, middlemen and farmers alike – since intensification is driven in part by economies and finances and, in part, by the expanding human population. We are now living in an age where people are significantly influencing the earth's resources, including animals. Policies that fail to acknowledge the tensions between economic growth and ecological integrity miss the point, suggesting a need to rethink the frameworks and rules that underpin social life. Different values and a more civic science, one that has a more significant participation in public policy processes, may be required to inform communities and enable society to make difficult decisions on complex, wicked and contentious issues like animal welfare.

Animal welfare, then, is not just about ensuring animals always have appropriate amounts of feed, water and shelter, are kept in good health, handled well, have opportunities for normal behaviour, or that any compromises to their well-being are justified and any harms minimized, though these are clearly very important. Animal welfare is also about whether modern culture, the systems driving, supporting and depending upon farm animals, can adequately and fairly provide for animals' needs. What changes are needed to ensure animal welfare, and farming, remain sustainable and in keeping with society's consensus view of what is acceptable? If, as is suggested, it is culture that is a driving force, then are we able to enhance animal welfare through cultural change?

The identity of a culture is found in its morality, religion, language, relationships, science, technology, education and so forth. The first step, it is suggested, is to acknowledge the complexity that is modern animal welfare (Chapter 1) and address the disconnect between people and animals (Chapter 5), between culture and biology, to resume the conversations started in modern times by Leopold, Potter and others, but arguably known in earlier times. Without them, the burden of animal welfare falls, arguably unfairly, on the people who almost exclusively care for animals, who can only reasonably make a living by intensifying animal production. As intensification may bring risks to animal welfare, society's increasing demand for continued, if not enhanced welfare, in the face of demands for more efficient and productive farming, becomes unsustainable. This reality must be confronted by looking to create the conditions in which animals can be farmed according to a consensus view of how they should be treated. Conditions that provide farmers with the resources, opportunities and confidence to farm to those standards since, after all, farmers must negotiate the barriers borne of being subordinate to the physical and social environments (Chapter 2). Whether the barriers are overcome accidentally, or in unplanned, organic, constructive, spontaneous or emergent ways, it probably does not matter but it probably will not best be by design (Ridley, 2015).

What we are talking about is the sustainability of a complex social-ecological system (Ostrom, 2009) of which animal welfare is but one part. The system is diverse and wicked, in part because it draws on the 'elite' for much of its functioning, and in part because it also supports that 'elite' providing food and a source of commerce. Understanding this system requires identifying the components and their place in complexity, a process complicated by the different languages of animal husbandry, animal welfare science, animal ethics, economics, policy and politics, and by the range of institutions involved including families, markets, agri-businesses, churches, sports clubs, pubs, governments, non-governmental organizations and so forth. Institutional and cultural factors contribute to our expectations of how animals should be treated, expectations that change as we adjust to different circumstances in time and space, contexts and beliefs. Earlier chapters have alluded to their diversity giving quite a context to the subject. In addition, we need to ask if all of us must be connected to our ecology, or can we support and trust others to do it on our behalf, to ease any guilt felt for the way animals are treated?

Humanity's prosperity is linked, at least in part, with farming. In a sense, the elite (Chapter 2) also includes farmers themselves (since they also benefit materially, at least in the western world). Importantly, this book is not a defence of farming, nor a plea on behalf of farming, but an acknowledgement that society is 'farming farmers' by providing the necessity to intensify, and the means of intensifying, production. As a source of food, commerce and many other things, farming is hugely important as evident over the last 10,000 or so years.

The belief in prosperity continuing to be dependent on a healthy farming industry is most obvious in the expectations that technology, smart-farming, precision agriculture, robotics and so forth will continue to drive more productive systems. At the same time, such systems need to ensure the welfare of animals is maintained, if not enhanced, communities maintained, and the impacts on the environment minimized, and past impacts mitigated. All this, while remaining profitable in the face of falling returns and increasingly being a social minority. Reflecting the disconnect between man and nature, animals and people (Chapter 5), the inconsistent expectation, sufficient and affordable food without affecting animals, communities or the environment, is perhaps most evident in the belief that farming industries must continue to shift their performance (Figure 6.2). This shift is essentially the story of farming: intensification is not without risks – unless you have the training and experience – 'the faster you go the bigger the mess'.

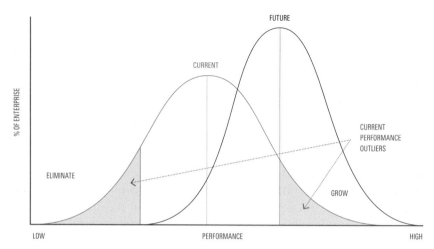

Figure 6.2 Societal expectations that prosperity depends on shifting the performance of most farmers, eliminating the poorer performers. 'The leading farmers of the future will year on year produce more while reducing the inputs into their systems' (*Courtesy Chartered Accountants Australia New Zealand*).

Consider, for example, the effects of domestication and intensification on animals. The more intensively animals are farmed, the more they are managed and selected for productivity and the more husbandry they generally require. For instance, 'less domesticated' Scottish Blackface ewes, a hardy hill breed of sheep that has a shorter and easier birth, are more attentive to their lambs immediately after birth, and require less human assistance (Dwyer and Lawrence, 2005). More intensively farmed animals can, therefore, require more husbandry, and this role has traditionally been what defines shepherds, a concern for those attempting to attract young people to farming industries. Similarly, the high-performing dairy cow producing large volumes of milk has been likened to a Ferrari. But Ferraris are fast, expensive and some, such as F1s, notoriously difficult to drive, requiring specialized skills and extensive team support. So it is with dairy cows; high-performing animals demand increased husbandry and veterinary skills. Good stockmen and women, veterinarians, nutritionists and other professionals are required to prevent and treat infertility, mastitis and lameness, the diseases of intensification. Additionally, some systems have reached their limits with greater stocking densities and production now counterproductive. Is it any wonder then that, and especially when combined with other stressors humans face (Chapter 1), systems occasionally fail and, like an F1 Ferrari, sometimes with catastrophic consequences? Frequently, society's response to poor animal welfare is to call for education or prosecution. There is no doubt there are many examples of poor animal welfare due to lack of caring, negligence, ignorance or even criminality (Chapter 1). However, society could also begin to more widely understand and acknowledge the drivers of intensification. Furthermore, that the continued transformation of the farming sector to realize the potential of agriculture for national economies requires a wide range of talents, skills and experiences that some individuals may not have. Is it fair to eliminate farmers who are not high performers, who will not go 'fast'? Should society assist 'slow' farmers? Do we need all farmers to go faster, and should either, or both groups, be supported if the benefits are to the wider community, in the form of a nation's economic prosperity? Should farmers have to use their land optimally to maximize production and wealth? As Wendell Berry (1984) entitled an essay – 'Whose Head is Using the Farmer and Whose Head Is the Farmer Using?' – there are many expectations of farming and farmers. For example, expectations that farmers need to 'adopt, understand, invest, communicate, collaborate, provide career paths, recognize, innovate, and do the right thing, etc.' are not uncommon. Is it fair to expect individuals to not only focus on their own interests, but also the wider community, without realistic reward? Does society adequately reward, even acknowledge, those farmers doing the right thing?

There is an 'inherent inconsistency' in the expectation that food is safe, sufficient and affordable, is produced without harming the environment or being cruel to animals or people. In addition, any costs in meeting these objectives are borne by farmers. But farmers also bear the burden, along with their animals, soils and environment. They are subordinate to the physical environment, especially the climate and the extremes of weather patterns, and to society (Chapter 2). And they must have the capacity or resilience to deal with volatility, both social (e.g. changing expectations of how animals should be treated such as greater use of pain relief for painful husbandry procedures) and financial (e.g. reduced income from downturns in international commodity prices). Collectively, this suggests not farming to the limits of the system if farmers and farm businesses are expected to absorb all or much of the risks. Yet intensification is a symptom of the need to farm more efficiently, to approach those limits, to make animals more productive.

What is good farming?

Agriculture has several broad economic, social and environmental goals. They include profitable, sustainable and environmentally safe production supplying people with products and opportunities for commercial gains (e.g. processing, distributing, marketing and retailing food, fibre and other products such as nutritionally adequate and safe food) in a fair and just or socially acceptable manner (Aiken, 1991; Zimdahl, 2002). The quest for economic prosperity through development and progress, science, technology and innovation, the work ethic, myths of the garden, capitalism etc., have all reinforced subordination by not questioning the long-term aims, just solving short-term problems by, for example, putting hens into cages, pigs into crates, or killing young surplus dairy calves. The lesson from intensive or factory farming is such systems are at odds with, or have become disconnected from, the expectations of the broader community. Science and technology, perhaps by default, provided the changes that, with the benefit of hindsight, should only really have been achieved with society's explicit approval.

Thinking like a mountain, then, is difficult as farming is constrained, not just by the physical environment, but by being subordinate to the elite, or the greater human population. As history has demonstrated, this is only a problem if it is not equitable and it is not equitable partly because the different parts of the system are not connected. Human culture is driving intensification but it has been unchecked. This reflects the view that the consumer, everyone beyond the farm gate, is king, wants and needs going largely unchecked as a result of being disconnected, perhaps alienated from farming. While on the one hand, it is becoming increasingly accepted that farmers require a social licence to operate, what sort of 'social licence' do retailers need to retail, consumers to consume? Consumers, all those beyond the farm gate, have a role in animal welfare, yet all too commonly we 'bay for tougher regulation, though not of ourselves' (Ferguson, 2012).

As elites, have we lost the connection and thus the respect for agriculture? As humans moved from living in bands to living in small villages and towns, kingdoms and empires, an economic system developed under the influence of trade and political relationships. Where once we gathered or hunted for a commodity (C), we have progressively moved to trading commodities and goods, that is, C→C and then via money (M), that is, C→M→C. So, in addition to not thinking like a mountain, or being constrained by or subordinate to society, the connection with biology (agriculture) has largely been severed. It is further disconnected or subordinated by a lack of moral perception; there is little recognition of animal welfare as a moral problem involving everyone, but it is seen only as the farmer's problem, or sometimes farm industries'.

Perhaps the main part of thinking like a mountain is being able to realize the extent of the problem, the context or narrative of farming. Farmers are at one end of this narrative, some either struggling, some perhaps not caring, the majority caring and doing well. At the other end are consumers, like kings, some caring, some not. How do we take responsibility between these extremes? It is suggested that it is the middle ground, that which we share and agree upon, that should play a greater role in determining the future of animal welfare, of the relationship between humans and animals. In her book *The Ethical Imagination*, Somerville (2006) suggests we tend to focus on things we disagree about, rather than finding the common ground or shared ethics. Like the Australians in the Great Depression who followed wallaby trails between waterholes, Somerville uses the term 'going on an ethical wallaby' to find moral resources we share, and are separated by, in our pluralistic, multicultural and global society. And like the Australians following wildlife trails, she believes that we also must recognize the importance of favouring the natural, of using different ways of knowing, and exercising restraint and accepting uncertainty, rather than seeking goals through force and domination. In this journey to find consensus in diversity and difference, 'all of us, not just the "chosen few," need to exercise our ethical imaginations' by integrating good facts (e.g. reason and science) and good ethics (e.g. imagination and creativity) into a seamless whole. In order to start from a position of agreement, the middle or common ground, Somerville suggests that our approach should be based on complexity thinking, identifying and dealing with issues using virtues such as trust, courage, compassion, generosity and hope. Any account of the future needs to focus on human flourishing, the promotion of virtues and encouraging people to develop or express them.

Farm animals are a resource for economies, humans making increasing demands, but the problem is it matters to animals how they are treated and the welfare of some is less than adequate. Furthermore, it matters differently to different people how animals are treated; farm animals exist or don't, suffer or don't and thrive or don't, because of the conditions people provide. This tension or dilemma reflects mixed ethical systems – commercial and guardian – requiring safeguards to ensure that important values of one are not cancelled out by the other. Reflecting on sustainable ecosystems, advocates of science observe that humans have become a major force of nature (Lubchenco, 1998; Röling, 2000). Since animals sustain human life, it follows humans are a major force, or driver, of animal welfare and that to address animal welfare, we must address human activity or deal with human behaviour. What Somerville (2006) is suggesting is that to reach a complex understanding of issues like animal welfare, we might have to 'consider notions of objective and subjective; the knowable and unknowable; the individual and the collective; duties to act and obligations to exercise restraint' in order to prevent, as some might fear, 'a takeover of ethics by economics' (Flynn, 2012). It is the thesis of this book that it is not just the behaviour of people directly involved with animals – the shepherds, farmers, cowboys, farmhands and so forth – but all of us, that must ensure that this does not happen. Currently, human activity is dominated by economic concerns, production and consumption and based upon this many of us enjoy good health and welfare and all the trappings of an affluent lifestyle. Little has changed since 1964 when Ruth Harrison's *Animal Machines* was published; we still have intensive or factory farms. We must do something different. We face a different challenge to the one science, technology and economics faced in having to feed the world after World War II.

Wendell Berry (1996), in *The Unsettling of America*, a review of the ideology of modern farming, lamented the loss of agriculture to modern agribusiness because of its impact on people, communities, land and animals, the cultural context of good farming. It is this deep, intimate, often tacit or ineffable knowledge that is threatened by the intensification favoured by an economic system that, seemingly almost exclusively, values production and profits, which society needs to evaluate. Berry also noted that the culture of agriculture is changing. It is like the rest of society, becoming very material whereas it is impossible for the material and spiritual to exist without the other. In other words, to preserve ourselves we need to respect and care for others, both the earth and its life.

Saul (2001) highlights, in *On Equilibrium*, that we have many ways of interacting with and knowing of the world. They include: (1) common sense or the shared knowledge borne of imagining the other, the whole and the relationship between humans and the environment. Common sense is not the exclusive realm of the expert or the individual, but the communal knowledge that ensures dominant beliefs, for example, that economics does not override other interests. (2) Ethics, or how to live within the context of the larger or public good: the compromises, often the costs, necessary for successful societal relationships. Legitimacy and leadership lie with citizens, rather than elites, and consideration of the context and reality of decisions is important. (3) Imagination, which enables us to be creative, and it is what makes us human. In contrast, imagination can be inhibited by excessive focus on technology and economics. (4) Intuition, the basis of actions we take without the luxury of careful consideration. Sometimes intuition is an expression of ineffable knowledge unable to be readily articulated, for example, stockmanship skills borne of a familiarity with animals and their needs. Intuition can also lead to beliefs that reason may confirm or refute. Finally, (5) memory or history, which shapes our thinking and our actions; and (6) reason, which has become dominant in the modern western world view. For instance, in relying heavily on science and evidence, there is a danger that the understanding of animal welfare is limited to performance, feelings and fitness, neglecting perspectives that recognize respect for animals, their dignity and even spirituality

(Chapter 1). Human nature is most effective if all the qualities of normal life are recognized, regarded of equal value and used in balance, or equilibrium, in our lives. As Saul has suggested, excessive reliance on reason, such as invoked by corporatism, represents the dehumanization of the individual citizen. While Berry lamented the loss of agriculture to modern agribusiness, Saul lamented the loss of all our ways of knowing to reason.

There is, then, a need to think about the future of animals, the environment, farming and people differently. Only then can we find, encourage and support ways of using science, technology and economics to shape and support that future. Sustainability requires collective thought and action – a widespread understanding – so that we can learn and adapt rather than benefit at the expense of one another, be they animals, the land, water, farm workers, immigrants, consumers and so forth. To do this, we will have to collectively understand, that is, connect all our well-beings with that of animals and the land. Or trust that someone has. How might we create or shape a world, or an economic system, which supports the world the way people collectively want it to be, be it material or spiritual, or both? Currently, animal welfare is a reflection or symptom, of a system, of farmers farming within the limits of an economically constrained or defined way of life, one designed or at least increasingly facilitated by science and technology, and our social institutions, based seemingly almost exclusively on the values of productivity and efficiency serving humans' material needs. Furthermore, the disconnect between most western peoples and farm animals, even nature, provides an opportunity for the relationship between animals and people to become dysfunctional, making it difficult to create or act sustainably or equitably towards something we do not know, that which we cannot 'see, feel, understand, love, or otherwise have faith in' as Leopold contended. Is the solution to reconnect humans with nature?

The importance of reconnecting is well recognized and there are potentially many ways of modern society connecting with farm animals. The earlier example of the UK school who took their pupils wildfowling (Chapter 5) is encountered widely. A popular experience for young schoolchildren is the visit of a farmer, a sheepdog and a handful of sheep, both species of animals well-habituated to the attention they receive. The show usually includes moving the sheep with the dog and shearing of the sheep. The show, exposing young children to farm animals at school, is enthralling, soliciting comments such as, 'Your show was a new experience for our children which they thoroughly enjoyed. It provided exceptional learning experiences which has reflected in subsequent class discussions, written language and art work' (Anonymous, undated, c). Such experiences also include the visits of schools to farms, the involvement of children in farm activities, and the interaction of children with animals at livestock shows.

The UK's Farm Animal Welfare Committee (2011) even goes as far as to recommend that schools make an effort to ensure that children learn where food comes from and that we have obligations to the animals that produce it. Farm visits and exposure to good husbandry are clearly an important part of that effort, as is the professional development of teachers to support that exposure, not only for children but also their families and wider communities. Or there's the introduction of farming into urban areas, reconnecting residents with the means of food production, such as the planned floating dairy farm in a city harbour (Boztas, 2016).

Secondly, there is the linking of consumers to farms through corporate social responsibility and product quality assurance schemes and independent verification of animal welfare claims (Chapter 4). Although there are several motivations for these initiatives, they do attempt to connect people (Figure 6.3) and are examples of the popular belief in the value of telling the 'story' of farming and good practices. These interventions, part of the animal welfare industry (Chapter 4), may be valuable in themselves, make people feel good and reflect expectations of care, but they do not address the underlying drivers that compromise animals and make it difficult to treat them humanely.

Thirdly, we can change the relationship between animals and humans. Take, for example, the selection of sheep requiring less shepherding (Chapter 5) resulting in an improved connection between human and animal needs. Similarly, better care of animals could be achieved by understanding and overcoming the barriers to farmers and others developing skills and establishing best practices by having access to authoritative advice and professional support, and, above all, providing fair reward and a sense of being valued.

Fourthly, rather than change the system in which we live, for example, expect people to treat their animals in certain ways, can we develop a

Figure 6.3 An example of a retailer connecting shoppers with the farmer (*Photo: Mark Fisher*).

system in which animals are treated in the way society believes they should be treated? For example, concern with the killing of surplus or unwanted dairy calves is, leaving aside any mistreatment (Fisher et al., 2017), in part because society does not put a value on the animal. Born to ensure the dairy cow produces milk, although a proportion are reared for future dairy cows, or for the beef cattle industry, many are killed for veal and ground beef, rennet used in making cheese, pet food, high-quality shoe leather, and new-born calf serum for use in tissue and cell cultures, vaccines, supplements and cosmetics (Trebilcock, 2017). While farm industries might be expected to make use of 'unwanted' calves, they are slaughtered because they are unwanted or not valued. Any alternative use or value depends on the ability of society to support it, for example, a demand for veal from older animals. This is a lack of moral perception; unwanted surplus calves are not recognized as a problem involving everyone, but it is seen as a farmer's or an industry's problem. Such a problem cannot be solved by farmers and their industries alone, but also by consumer demand for whatever product an alternative use produces; thus when society values the young calf, it will not have to be killed at such a very young age.

Fifthly, and related to the previous example, perhaps there is an opportunity to learn about and manage humanity's collective behaviour so that biology or ecology guides it, rather than a more dominant focus on material consumption and economic returns. An emphasis at animal, farm or even industry level obscures the larger social networks and institutions they are part of. Is there an opportunity to provide society with leadership by considering shaping and changing the behaviour of people benefiting from animals, as well as the behaviour of people caring for them?

The western world view, reflecting our educated, industrial, rich and democratic culture, has been spectacularly successful but it does not mean it is the only way, or the best way for the future. Neither does it mean it is not without its problems as threats like inequality, terrorism and climate change attest. Perhaps the diverse cultures of the world provide other insights into thriving in a sustainable way (Davis, 2009). As noted in Chapter 1, much of this book, and our understanding of animal welfare, is based on the perspectives of those in Europe, North America and Australasia, countries described as largely Western, Educated, Industrialized, Rich and Democratic (better known by the acronym WEIRD). This stance not only reflects the author's experiences with farming, science, ethics and animals, but it also mirrors the development of modern intensive or factory farming and the fact that 90% of the population can spend their energy on activities other than producing food. Although other peoples and cultures have, or are developing, such farming systems, what insights do people from different communities, that is, non-western, non-educated, non-industrialized, non-rich and non-democratic societies, provide? The possibility that WEIRD societies were unusual and not representative of all humans was raised in human psychology and behaviour (Henrich et al., 2010a,b). Caution was urged in basing findings about human nature on studies drawn from such a relatively small and unusual, even extreme, segment of humanity (96% of subjects came from 12% of the world's population, many also being university undergraduates). For instance, most urban children know only one animal, *Homo sapiens*, so any conclusions from subjects reared in such impoverished and unnatural environments must remain tentative. People with a greater affinity with the natural world prefer to use inferences from folk-biological knowledge that considers ecological context and relationship among species, despite having similar reasoning to WEIRD subjects. And while typical western subjects rely on the moral principles of justice and harm/care, others rely on them but also on a wider range of principles including autonomy, community and divinity. There are perhaps no societies as deeply alienated or disconnected from other life forms as those shaping the modern, western, affluent relationship between humans and farm animals (Nelson, 1993; Wilson, 2001).

Nature and culture are woven together in other cultures and non-WEIRD communities might provide guidance in their relationships between animals and people. While hunters accumulated knowledge through patient observation and close interaction with the natural world, our modern educational systems have specialized people within institutions, separated from the natural world apart from field-trips where they effectively view nature as tourists. Furthermore, as technology becomes more sophisticated and complex (e.g. electronic fish-finders and infrared rifle markers) the importance of knowledge of hunting borne of intimacy of the natural world over many generations is lessened. The opportunity to learn from animals is also vastly reduced. In some societies (Nelson, 1993), 'luck' in hunting is attributed to an animal's spiritual power providing a source of restraint based on respect for the animal and humility. This core human value promoted a more sustainable relationship with the natural world. In contrast, the natural world is unknown to many people in WEIRD societies, partly because they are urban and partly because their world view is based in science, philosophy and religion, especially empirical knowledge. WEIRD humans might not understand or experience the natural world as hunters might have, and some people still do. For example, some peoples see themselves as being intrinsically linked to the natural world, giving a different understanding to the connections and relationships in the world, and the interactions between the spiritual and physical worlds. Thus, there may be value in at least acknowledging that other ways of understanding the world exist and exploring what they might add to our own relatively unrestrained relationships with animals and the natural world (Nelson, 1993). Although non-WEIRD and indigenous cultures have changed with time, many embracing western values and world views, and no doubt contain, like any people, their share of 'puritans and sinners, conformists and lawbreakers, and all shades between', their ideal of respect for the animal, something akin to stockmanship, may provide a path for the future.

To return to killing, some believe that the best western standards, inducing insensibility before killing (often, at least commercially, electrical stunning followed by throat-cutting or sticking but sometimes by gunshot or knife), are inhumane, at least compared with khoj özeeri (Rymer, 2012). The Tuvan people of southern Siberia are historically nomadic herders, moving their aal – an encampment of yurts or tents – and their sheep, cows and reindeer from pasture to pasture as the seasons progress. If the method of slaughtering livestock is part of humans' relationship with animals, khoj özeeri is unusually intimate. Through an incision in the sheep's hide, the slaughterer severs a vital artery with his fingers, allowing the animal to die without alarm. In the language of the Tuvan people, khoj özeeri means not only slaughter but also kindness, humaneness, a ceremony by which a family can kill and butcher a sheep. Khoj özeeri implies a relationship to animals that is also a measure of a people's character: 'If a Tuvan killed an animal the way they do in other places' – by means of a gun or knife – 'they'd be arrested for brutality.' Apparently, even the reputation of Genghis Khan's army was that they shed the blood of humans more readily than they did the blood of sheep (Cherfas, 2012).

A final example, of the complex interaction or relationship between ourselves and animals, is the way animals are perceived. The stupid, harmless or boring perception of sheep in western culture may be an artefact of the western belief in valuing and associating individualism with intelligence, originality and leadership. In cultures valuing conformity (anyone can be an individual but the ability to fit in is a social skill), such as China, the sheep is considered highly intelligent (see Franklin, 2007).

As many have articulated, there is concern with the care of animals in modern intensive systems, a reflection of narrowly focussed values of efficiency and productivity in the western world in which the quality of human life is defined, almost exclusively, by material and consumptive concerns. These values and world view may reflect, in part, the lack of contact people have with animals and the land and, in part, the benefits and affluence borne of that disconnection. By specializing in increased production, and being able to transport it, we have created a modern world, albeit one that has lost the more holistic world view more characteristic of hunter-gatherer and other societies. Most of us live in highly advanced isolation having lost touch with the lessons of living with animals and the experience of interdependence and community. The farmer, once a stockman or steward of the land, its plants and animals, is increasingly concerned with financial resources and constraints (Hodges, 2000). In Japan, experience with running a business is considered an essential part of good stockmanship (Tsushima, 2007).

What can we learn from these non-WEIRD societies? Perhaps that we are all part of a greater whole. Perhaps we need to look at traditions far removed from our own, those less alienated from animals and farming. Perhaps pastoralism and extensive farming are as far removed as we can get today. So, if we are not connected, and other societies are or once were, how can we reconnect? And finally, what can we learn from the non-human world? Humans focus almost exclusively on using the senses of sight and sound, separated from much of the rest of life; perhaps our other senses would help in overcoming the disconnect (Wilson, 2014).

Modern sophisticated connections

If the sentiments in this book are correct, and modern societies are disconnected from their biological and ecological foundations, we appear to be living in a paradoxical age. Modern transport from container ships to jet aircraft, communications like the internet and smartphones, and political activities facilitating a global marketplace suggest we are more connected than ever. However, the connections we do have are individual and focussed, and without the community and depth of knowledge of a full understanding. Few of the modern connections appear to be serving understanding of the natural world, including animals, in any sophisticated way. This has been more eloquently put by the Irish poet and playwright Oscar Wilde (1854–1900) who, anticipating his release from prison and spending time near the sea, wrote, 'It seems to me that we all look at Nature too much, and live with her too little.' Or, more poignantly, Alexander von Humboldt (1769–1859), a German naturalist and explorer who observed that, 'The most dangerous worldviews are the worldviews of those who have never viewed the world.' Animal welfare is driven by human activity, both suffering and demands for humane practices. Improving animal welfare then requires a change in human behaviour; the answer lies not just in science, husbandry and ethics but in seeing and acknowledging what is causing the problem and what needs to be done to fix it (Röling, 2000).

An essential step in connecting is to understand the system: Wilde's Nature, von Humboldt's world views, and the societal and political institutions influencing and facilitating farming and animal welfare. Animal welfare is part of a complex system of networks, connections and dependencies. It is about sun and soil, genes and behaviour, husbandry and production, ethics and values, markets and wants, perceptions and prejudices, not just those of the farmer, or even the farmer and consumer, but all of us. It is this complexity that we must embrace by looking at the whole system and, however we can, balancing it against the increasing specialization of civilization, a process begun in the Stone Age. In explaining how the new era of globalization has become dominant and how it shapes politics, commerce, the environment and relationships, journalist Thomas Friedman (1999) in *The Lexus and the Olive Tree* (metaphors for the struggle in balancing modernity and tradition) stated that the best way to see complex systems is often through simple stories conveying complex truths, rather than theories. In the rush of food from the farm to the plate, stories have been ignored, land and food have become commodities. Furthermore, in a world valuing specialization, people, students, professors, journalists, scientists and others, who understand complex interactions and can think 'like a mountain', are also needed. Friedman draws on American theoretical physicist and Nobel laureate Murray Gell-Mann's *The Quark and the Jaguar*, advocating that we also look at the whole, discussion of the big picture unfortunately all too often relegated to cocktail party conversation. It is this big picture of the relationship between people, animals and nature that is at the heart of ecological and social or cultural systems (Pretty, 2007).

The whole, or big picture, is not simply about reconnecting farmers and consumers, but all of us as the diversity of interested parties in Chapter 5 indicates. And we must reconnect in different ways, through memories, stories and places, local conditions, connections that give us understanding and knowledge. If we wish for animal welfare and economics to exist together then perhaps we need to create new values or incorporate them into a variety of existing values. Searching for the common ground and shared understanding, as advocated by Somerville (2006), might begin by considering the animal welfare system.

Living with nature and viewing the world could begin by considering animal welfare as a system involving everyone, since animal welfare is increasingly everyone's business. One simple depiction has animals at the centre (Figure 6.4). Arranged in this way, the system acknowledges that each group has a role, and thus a responsibility, for animal welfare. The schematic design also provides an opportunity to see, and question, some of the features of the system. Firstly, costs and benefits tend to be borne differently. The benefits from animal use tend to extend outwards, while expectations for the care of animals tend to be directed more towards the centre. For example, there is an increasing expectation that farm animal

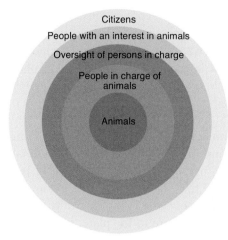

Figure 6.4 A schematic representation of the animal welfare system arranged as a series of bands of people surrounding farm animals. The system includes farmers or persons in charge of farm animals, those with oversight of persons in charge of animals (e.g. animal welfare inspectors), those with an interest in animals (e.g. consumers, retailers, animal advocates, industry organizations and professional associations) and finally citizens, who, while not necessarily having direct vested interests in animals, have a special role in the public good or democratic process (from Fisher et al., 2014).

welfare should be audited, verified or monitored. However, should not other parts of the system also be open to scrutiny where their activities impact on animal welfare, for example, retailers, consumers and even animal advocates? Without acknowledging others' responsibilities, are not farmers scapegoats, having to justify what many others benefit from? A second feature of the system is that, like tourists, individuals within each of the bands see the world from their own perspectives, in varied but often simplistic ways.

No single group has exclusive possession of the truth, but see the world borne of their upbringings, education, experiences and aspirations for the future. A good visual example is the size of the lion in many of the 15th-century portraits of St Jerome (see Figure 6.5). The story goes that the 4th-century saint reputedly befriended a lion by removing a thorn from his paw. In many of these works, the lion is portrayed as something less than the king of the beasts and, in some, more like a dog with tail, mane, feet, face and ears all quite un-lion-like. Regan (1983) suggested the lion was painted from imagination and 'scant information and anecdotal tales about lions', artists having never seen the animal (in contrast, more recent artists, like Jean-Léon Gérome (1824–1904), have produced depictions that are very lion-like). It is a reminder that we, all of us, need to understand the animal. The lack of knowledge of animals is still apparent. An image, promoting a sausage-making company, of a pig 'bathing in mud' was regarded as 'disgusting' and 'morbid' (Mortimer, 2018) despite alternative views, reflecting natural behaviour, that 'happy pigs are dirty!' (Lassen et al., 2006b).

To incorporate the valid viewpoints of different parties, dialogue or true understanding must be encouraged and facilitated so that decisions are

Figure 6.5 *Saint Jerome Extracting a Thorn from a Lion's Paw* by Master of the Murano Gradual (d. 1473) from the second quarter of the 15th century (*The J. Paul Getty Museum, Los Angeles, courtesy of the Getty's Open Content Program*). (Available in colour in the plate section between pages 178–179.)

based on varied understandings, of experts such as scientists and veterinarians, those endowed with practical knowledge such as farmers and stockmen, and those with an interest in animals such as animal advocates and others. It is not just from this group, but society's collective wisdom, that decisions about compromises are made. Democracy depends on openness, reliability, appropriateness, responsiveness and two-way communication (Gore, 2007), and a means of making intelligent decisions and uncovering different perspectives. Those in positions of authority in animal welfare, by their election or employment, must think on behalf of others and act for those who have entrusted authority to them. There must be acknowledgement that people have different views, those views must be acknowledged, improved upon if necessary, and if they cannot be accommodated then it must be explained why. Learning more of the features and expectations of others in different bands may act to change or reinforce our responses.

Finally, given the number of different bands of people, and their diversity, and the dilemmas associated with using farm animals, about which reasonable people can disagree, how good is society at acknowledging those different roles and responsibilities? How well do we give each other the confidence, resources and opportunities to undertake our respective roles? How well do we empower each other to produce an equitable system? Thinking in terms of the system, rather than as particular individuals or groups with personal or institutional interests and motives, requires cognizance of the whole system. This, combined with the systematic reflection of issues in the public sphere as a genuine and comprehensive concern for the interests and well-being of those who entrust the individuals and groups with their respective roles and responsibilities (Thompson, 1999b), may be the only real alternative to the combative nature of public engagement where participants can be viewed as 'tourists' or 'scapegoats'.

Farm animal welfare is society's collective responsibility, from farmers to citizens, and it is necessary to acknowledge and understand the system. Building connections is not just about telling, but two-way communication or dialogue, shared understanding (Bohm, 1996). In other words, to act in the best interest of others through focussing on shared values, or the middle ground, before working on those representing or reflecting disconnections. A system such as that represented in Figure 6.4 does not provide guidance on how animals should be treated and therefore needs principles to help guide actions (e.g. Table 6.1).

These principles, if correct, should reflect what it is to be a good human in a flourishing civilization, bringing together diverse expectations and realities by acknowledging valid concerns, providing information and involving people (Fisher, 2010; Korthals, 2008). The aim should be to acknowledge the true costs of human–animal relationships fairly and openly, balancing our needs with those of animals. This is important for all of us, as sharing

Table 6.1 *A suggested set of principles describing and helping guide the animal welfare system.*

Principle and examples

(1) Human well-being is dependent on animals
- Animals are an integral part of our lives, both material (economic) and spiritual (cultural)
- Cannot live without incurring some debt to animals
- Use of animals is permissible subject to society's approval

(2) Caring for animals is part of being human
- It matters to people how animals are treated
- Care of animals relates to ensuring their physical, health and behavioural needs are provided for
- Care not only includes preventing or alleviating suffering and enabling pleasure, but respecting their natures and treating them with dignity
- Reflects the co-evolution of humans and animals and the skills of stockmanship and animal husbandry
- Society should ensure those responsible for the care of animals have the time, resources and confidence to do so

(3) Animal welfare is part of a complex
- Animal welfare has cultural, social, environmental, political, ethical, religious and other dimensions
- Farming also has several dimensions including profitable, safe and sustainable production of products to the satisfaction of human needs in a socially acceptable manner
- Animals are valued for their own sake and not just because they're sentient or of use to humans
- Concern is not limited to just the well-being of individuals but also includes the future of various strains, breeds and species

(4) Justification requires appropriate knowledge and understanding from a range of disciplines
- Animal husbandry, scientific knowledge and available technology
- Common sense, ethics, imagination, intuition, memory and reason
- Common morality and the insights of other ethical theories

(5) Many beliefs guide the use of animals
- It is acceptable to use animals for human benefit providing that that use is humane
- Some harms are prohibited, other harms are justified by the benefits they bring but those harms are required to be minimized
- While the interests of people can be differentiated from the interests of animals, so too can the interests of animals in close relationships with people (e.g. pets compared with wildlife)

(6) Oversight belongs to all
- Transparency and shared understanding are required by those to whom animal welfare is entrusted, including the context, drivers and barriers around compromises, as well as the costs and benefits of those compromises
- Citizens should be encouraged or empowered to participate
- Assumptions and worldviews should be critiqued, and alternative and diverse views considered

(7) Consensus
- Acceptable treatment of animals reflects a societal consensus, i.e. the middle ground
- Other perspectives should be understood, respected and considered, even if they cannot be reconciled
- Reasons for stances should be articulated to facilitate understanding and progress

---

resources, both the benefits and the burdens, needs to overcome any inequity in the system. Farmers may need more certainty in decisions if they are to invest what are increasingly larger amounts of capital, as might consumers if it means the end of the era of cheap food and prolific commerce, something many affluent societies may not be willing to renounce since 'greed doesn't generally bother the greedy' (Gaarder, 2015).

What then is good agriculture? That which supports and promotes human health, just communities, and the natural world, or that which produces abundant and inexpensive food, and provides financial prosperity? Mechanisms must be created to balance interests and resolve conflicts, based on a more nuanced understanding of their complexity and the interdependence of animals and people.

One such mechanism is to reduce consumption, for example, meatless Mondays (de Bakker and Dagevos, 2012). Similarly, there was the 1940s 'war on want' that used the opportunity of World War II to engineer social progress in the UK by attacking excessive want, along with disease, ignorance, squalor and idleness (Beveridge, 1942). However, focussing on efficiency and productivity supports consumption, the focus of our economic systems. What human–animal relationships will achieve new, more equitable and prosperous societies? And what systems are needed to support them? Answering these questions will require encouraging critical reflection and investigation, creating opportunities for different perspectives both in farming and the food supply chain and, importantly, in political control.

Another mechanism is to reward farmers for doing the right things, producing food by means that individuals and society judge to be acceptable. This may require being able to distinguish between products and paying a premium. While some producers and retailers favour higher standards of animal welfare to ensure their market share, competitive advantage or corporate responsibility, it is unclear whether consumers can sufficiently identify a credence attribute. Nor is it clear if farmers are rewarded, even recompensed, for superior production systems and costs. While growing numbers of people may be translating interest in animal welfare into purchasing intentions, there is much to understand and address in the way of their needs and expectations for presentation and labelling. It is probably unrealistic to expect behavioural changes related to the provision of information since it is sometimes rational to be ignorant, and positive depictions are relatively more difficult to portray, work more slowly and fade away more rapidly, than the impacts of negative depictions of animal welfare (Verbeke, 2009).

If it is not the producer or the consumer who can adequately address animal welfare, then we are left with citizens. Good citizens are those that pay their fair share, know of society's needs, are vigilant stewards for future generations and are compassionate, even outspoken, nonconforming and disinterested (Saul, 1995; Sachs, 2012). To not at least attempt to better understand these challenging perspectives is to risk not being good citizens.

However, modern society is characterized by the decline of civic virtue and of social responsibility. We are too distracted (Sachs, 2012) or perhaps distrustful of politicians to maintain the habits of an effective citizenry. That role increasingly rests with interest groups. It is suggested that the balance as individuals, consumers, citizens and members of society is reclaimed by recognizing our dual roles, both as individuals and as members of society both responsible to and sharing values with others. Citizens can be advised by experts and interest groups but must retain the power to control and shape the future.

Changing the way we think is at the heart of things. If we see the whole picture our thinking might change and to do so may require reconnecting as a society or community, that is, collectively, with our biology, through our values, by expanding democratic deliberation or citizenship. To paraphrase what Gundersen (1995) noted regarding the environment, many scholars have failed to appreciate the complexity of animal welfare so that a more collective, holistic and long-term view guides it, and especially an equitable one. To think like animals is our challenge, to understand them in their own terms, as well as ours, empathy, for both animals and other people, being central to understanding.

If farm animal welfare, and farming, is a shared obligation then it requires economic, educational and political solutions to achieve an equitable and sustainable future. How do citizens contribute to or take that collective responsibility? While individuals' and communities' attitudes, beliefs and values influence and determine the way animals are treated, there are two general but enmeshed ways in which animal welfare is determined by society. The first is public, through the law, and the second is private, through the marketplace. The two are linked, requiring solutions as diverse as 'changing minds and moving markets' with economists, sociologists and psychologists required as much as, if not more than, traditional scientists, veterinarians and animal advocates in addressing animal welfare (Sandøe and Jensen, 2013). For example, a real solution to the unwanted killing of young, surplus dairy calves might be to develop and support a market for veal based on those animals having a good life and a humane death. The development of a farming system, for example, rose veal, may require redesigning current farming systems, for example, aligning additional animals with pasture availability, and with slaughter and processing requirements, but, and significantly, also supported by consumer demands making it chic to eat veal, just like it once was to eat the rabbit. Such cultural transformations will be at the discretion of the individual, albeit facilitated by societal incentives. Solutions to complex problems (those that are distributed across a range of players, are difficult to predict and involve conflicting goals) require drawing various insights, not just those able to be provided by science, important as it is in the face of polemic rhetoric and emotive opinions, but also ethics, philosophy, social science, sociology, economics etc. It has been suggested that the key to taking responsibility for complexity includes (Jones, 2011; Yeates et al., 2011): understanding animal welfare as a complex problem, recognizing the constraints and opportunities; giving people more autonomy through engaging local institutions; building trust with stakeholders; taking accountability for learning; and broadening dialogues.

These mechanisms have at their core a philosophy of empowering people to be responsible for their own actions or the systems in which they live and work, of allowing the solutions to emerge from communities rather than imposing them. The role for citizens is in determining what is acceptable, even interpreting the legislation, and taking responsibility, since in a strong society the state and its institutions are firmly attached to its people. To complement the consumer king, do we need the consumer citizen? At present, the conflict between animal use and human benefits is played out in the conflict between, for example, institutions, stakeholders, governments, animal advocates, media and individuals, and between the ethical, economic and other ideologies. Ethical citizenship, cognizant of the connections and complexities inherent in animal welfare, may be a way for the community to ensure animal welfare is sustainable. If we do not do it ourselves, then we have to trust that something or someone does it on our behalf, not vested interests but disinterested and trusted. Those given the responsibility should be supported by a public media prepared to critique and be open about what grounds they have undertaken that critique, and to use tools such as doubt and scepticism, examining the beliefs and the believers, especially those in positions of power and authority (Sim, 2006). Many people might not have the resources, opportunities or inclination to contribute as citizens, suggesting there is a place in society for the likes of advisory groups acting on behalf of citizens. Perhaps the most significant opportunity to contribute to unravelling the dilemmas associated with benefitting from harming animals is challenging people as citizens, not just as members of interest groups, to understand and consider those common expectations. In *The Wisdom of Crowds*, Surowiecki (2004) holds that many people, with diverse opinions, provide a more complete picture of the world than do expert advisory bodies. Furthermore, if expert groups cannot be trusted, then collating self-interested perspectives makes good sense, especially since complex problems require multiple perspectives. The more important the decision, the more important it is that many people make it. In contrast, the dangers small groups face is consensus, the illusion of certainty over the reality of doubt. As Patrick Nairne (1921–2013) of the UK's Nuffield Council on Bioethics stated, 'bioethics is too important to be left to scientists, doctors and academics. It's on the agenda for all of us' (Vines, 1994). Complex ethical issues, like animal welfare, cannot be left to interest groups, scientific, advocacy or political. Questions citizens might ask include: is animal welfare all that matters in considering our relationship with animals? Should 'public' concerns transform the relationship with farm animals? Are there limits to improving the productivity and efficiency of animal farming? And if we reach the limits of the animals, the farmers and the environment, how does, or should, the economic system respond? It is suggested that some intensive poultry, pig and dairy systems are reaching or have surpassed some of those limits already. And can we, all of us, live a good life and have animals also live a good life? What is reasonable and necessary use of animals that balances the reasonable and necessary compromises to animals? How much does human consumption drive animal welfare?

These are deep and profound issues possibly requiring changes in human lifestyles, and in production and consumption. In reflecting on a similarly wicked issue, climate change, Sir Geoffrey Palmer (2015), former Prime Minister of NZ, held that leadership is required to transform the economy. Leadership and decision-making need to reflect not just the opinions and preferences of interest groups, political considerations and public opinion, as modern politics seems to be increasingly susceptible to, but evidence aligned with equitable human needs. If humans are to get to the year 3000, as Potter (1990) reflected upon, we have to think like the mountain. Rather than animal welfare, perhaps we should focus on animal farming, even agriculture, and let it be informed by animal welfare, food safety and security, the environment and human flourishing in general. A narrow scope, such as animal welfare, especially where the economic benefits to mankind seem to be the dominant justification for compromises to animals, obscures the complexity.

Perhaps there is an alternative, to accept that some live less than a good life, just as the people of Omelas did. If Omelas represents the relationship we want with at least some farm animals, then we will have to develop the psychological and institutional mechanisms to not have to 'walk away' like some of the people of Omelas. However, rarely is walking away from a relationship, such as that we have with farm animals, a reasonable way of dealing with it.

A new revolution?

The narrative of our relationship with farm animals is often thought of as a series of revolutions. The Neolithic Revolution brought a change from the hunter-gatherer relationship with prey animals and ushered in the beginnings of farming and animal domestication. The Agricultural Revolution saw new methods of farming result in increased production. Finally, the Industrial Revolution, accompanied by advances in science and technology, has brought us to our modern world, where in some systems, the spirit of the relationship between humans and animals is questioned by many. Throughout the more than ten millennia of changing relationships, both animals and humans have changed, the most remarkable changes being in the way humans live: from living in bands of largely foragers to living in a largely urban, global and complex civilization somewhat disconnected from its biological foundation. However, parts of the modern world appear unsustainable, including our concerns with or responses to animal welfare. Increasing intensification of farming, along with increased constraints, exposés and regulations, is not sustainable. Farming, along with other developments like transport, science and medicine, as well as the notions of competition, property, work and consumption (Ferguson, 2011), has brought civilization and mankind its affluence, and that affluence has resulted in questioning of farming systems. It is suggested that the next revolution will need to focus on people (Figure 6.6).

Figure 6.6 The role of people in animal welfare: the next revolution? While farmers (*Photo: Ross Buscke*) are directly responsible for the animals (*Photo: Mark Fisher*) and are the most important determinant of their welfare, people (*Photo: European Commission*) also have an indirect role and thus responsibility. (Available in colour in the plate section between pages 178–179.)

We need to understand an animal's welfare not just as what the animal feels or experiences, but as what both animals and people feel and experience within a modern, sophisticated and connected system supporting farms, animals and ecosystems, whatever they might be. That connection might require creating an economic system that can reward people for caring for animals the way society deems acceptable and that takes responsibility for providing those rewards to ensure the relationships between farm animals and humans can flourish. There are usually three broad arguments for improving farm animal welfare: pay a more realistic price for produce, reduce consumption (but that reduces the economy) or redesign the system. Deciding what to do will require acknowledging and addressing the underlying drivers of animal welfare, the ultimate as well as the proximate causes of poor animal welfare, embracing complexity rather than seeing animal welfare from a position of unfamiliarity and simplicity, and managing rather than resolving issues, as is characteristic of wicked problems. Whatever approach, it will need to ensure those with the responsibility for farming animals have the resources, confidence and abilities to care for them according to society's demands.

Animal welfare is wicked in part because, in attempting to shape the future, we begin by failing to understand the present, or prefer not to dwell on it. To paraphrase David Suzuki, the Canadian scientist and environmental advocate commenting on the environment, 'the difficulties we're encountering in solving our animal welfare problems are not scientific or technical; they're social ... we already know what must be done and even how to do it; we're just not acting' (Suzuki and Taylor, 2009). It is our social, economic and political choices, choices based on knowledge and values, that have resulted in the farming systems we have. As the wildlife slogan says, 'remember, when the buying stops, then the killing can too.' There is a need for genuine understanding; what are the issues, what information do we need to consider them, and how can we involve people to overcome simplistic portrayals of the complexity of animal welfare?

In Chapter 1, two approaches to animal welfare were introduced: fitting animals to farms and fitting farms to animals. A third way might be envisaged: fitting people to animals and farms. The expectations of people could align more closely with what animals, farmers and others can reasonably provide. But the connection must be of both individuals and communities, and both social and ecological. At present, animal welfare is most importantly, or so it seems, a propaganda battle leaving people concerned but confused. This is especially true of advocacy stances, rather than providing information allowing people to make up their own minds. Differences between expectations and realities need to be brought together, acknowledging valid concerns, providing information and involving people. The 'battle' is made more complex by treating farming as an aggregate. It is too general to draw conclusions, denying problems exist and proposing simplistic analyses of complex issues; we must avoid advocacy and provide knowledgeable research and analysis of the issues (Fraser, 2001). And, as the philosopher of science Sir Karl Popper (1902–1994) maintained, we should identify and improve arguments before beginning to criticize them. A first step might be to recognize society's dependence on animals, portray those issues driven by intensification and, where necessary, find more equitable solutions by creating the conditions enabling farmers to farm the way society wants them to farm. If the most powerful driver is economic, then create the economic conditions that incentivize or reward the animal welfare systems that society wants or values.

Modern society's farm animal welfare is largely the result of the system in which animals, farmers and others find themselves. We must think about that system if we are going to accept or change animal welfare. If society is to prefer pastoralism, extensive pastoral farming, or any similar such system reflecting traditional perspectives of farming, then society must reward and support it. If it is to be intensive, factory-farming, then society must live with it, as the people of Omelas do.

The institutions responsible for animal welfare generally operate around laws and economics; perhaps they also need to look at people, and why they behave in the way they do, by understanding the context and the complexity of the situations that influence animal welfare, as well as what the animals might be experiencing. Do we need to think about people as well as animals in understanding animal welfare, as people are generally acting for good reasons? Ensuring people have a good connection and understanding of the extent, nature and complexity of the animal welfare system is crucial. Alienating people, be they some farmers who have neglected to care for animals, or some activists who have extreme expectations of their use, is not a sustainable future path. As animal welfare reflects a system that includes all of us, perhaps we must acknowledge that we, all of us, need to value animals more and be prepared to take responsibility for any improvements we want, be they the costs of greater space, slower growth, assurance schemes or verification of compliance with standards. Animal welfare, then, can be addressed at the animal level and at a societal level and both farmers and society will have responsibilities. Society's key role could well be in valuing animals commensurate with its expectations of how they are to be farmed, that is, providing rewards, incentives, support or financial encouragement enabling farmers to husband animals according to society's expectations.

A true understanding of the relationship between humans and farm animals and, more importantly, an equitable one, may require articulating the tensions and the underlying reasons for the current differences between expectations and reality. Perhaps it will also require different values connecting people, animals and the environment. Traditional morality, the common ethical theories described in Chapter 4, attempts to deal with the relationships between people and how they share resources. Perhaps the future will require new forms of ethical theories, ones based on what relationships we want with the natural world, rather than principally on how we want to interact with others. This may require a forum for encouraging and exploring options, and a source of critiquing them, rather than the more usual reliance on authorities reflecting their interests (Sim, 2006). In articulating the complexity and tensions, and the underlying reasons for them, new forms of engagement and means of dealing with them may be required. For example, although increasing animal productivity and enhancing animal welfare are both valued, they can be inversely related. To maintain a desired level of animal welfare then, society may be required to support, reward or incentivize systems having less productivity, or accept highly productive systems with less than optimal welfare. This is a political decision, suggesting the future of animal welfare will have to draw more on subjects like political philosophy, how society should be arranged and our responsibilities within it. Reasonable people can disagree about what is fair and acceptable, and about how resources are used and distributed, the domain of ethics and politics. In this, animal welfare is only one part; food security, the environment, social aspects and so forth are also important. Individuals, be they farmers, transport operators, food processors or whatever, need to be able to afford to behave in ways that act in society's best interests (Harman, 1976).

Some of the necessary changes to the relationship between animals, people and the environment might be achieved, others probably will not. I am reminded of McInerney's (2013) postscript to a paper given to a meeting considering responsibilities to animals primarily in our care:

There is a tendency for international conferences to be rather like multinational bombing raids. The various participants fly in from different directions at varying heights and speeds, drop their particular ordnance on what they believe is the target and then fly out again, having taken it all very seriously and professionally and believing they have done a good job. Whether the target was destroyed as intended, or just a big mess made on the ground, is inevitably uncertain and something that only becomes evident when the dust has cleared. Everyone feels it was an important thing to have been part of, however.

The challenge, our challenge, your challenge, is to acknowledge our role in animal welfare. We, all of us in society, are part of the problem and therefore part of the solution, however difficult that may or may not be. If we do not, are we too just like those bombing raids? The final chapter considers the relationship between humans and farm animals, our place in the natural world.

Further reading

In Search of Civilisation. Remaking a Tarnished Idea. **John Armstrong**. Allen Lane, London, UK, 2009.

The Unsettling of America. Culture and Agriculture, 3rd edn. **Wendell Berry**. Sierra Club, San Francisco, CA, 1996.

The Wayfinders. Why Ancient Wisdom Matters in the Modern World. **Wade Davis**. Anansi, Toronto, Canada, 2009.

Prosperity without Growth. Economics for a Finite Planet. **Tim Jackson**. Earthscan, Abingdon, UK, 2009.

A Sand County Almanac. And Sketches Here and There. **Aldo Leopold**. Oxford University Press, New York, NY, 1949.

Out of the Wreckage. A New Politics for an Age of Crisis. **George Monbiot**. Verso, London, UK, 2017.

Bioethics. Bridge to the Future. **Van Rensselaer Potter**. Prentice-Hall, Englewood Cliffs, NJ, 1971.

Global Bioethics. Building on the Leopold Legacy. **Van Rensselaer Potter**. Michigan State, East Lansing, MI, 1988.

Agri-Culture. Reconnecting People, Land and Nature. **Jules Pretty**. Earthscan, London, UK, 2002.

The Earth Only Endures. On Reconnecting with Nature and Our Place in It. **Jules Pretty**. Earthscan, London, UK, 2007.

The Price of Civilization. **Jeffrey Sachs**. Vintage, London, UK, 2012.

The Unconscious Civilisation. **John Ralston Saul**. Penguin, Ringwood, Australia, 1995.

On Equilibrium. **John Ralston Saul**. Penguin, Ringwood, Australia, 2001.

The Ethical Imagination. Journeys of the Human Spirit. **Margaret Somerville**. Melbourne University Press, Carlton, Australia, 2006.

The Wisdom of Crowds. Why the Many Are Smarter than the Few. **James Surowiecki**. Abacus, London, UK, 2004.

Six Steps Back to the Land. Why We Need Small Mixed Farms and Millions More Farmers. **Colin Tudge**. Green, Cambridge, UK, 2016.

CHAPTER SEVEN

The fall and rise of the hunter-gatherer

A representation of the modern relationship between humans and farm animals is that depicted in Figure 7.1 by the German artist Max Ernst (1891–1976). *City with Animals (Stadt mit Tieren)* is an example of expressionism, the artist portraying his emotions rather than any accurate external reality. As with most people, the relationship between humans and animals is dominated by emotions. In addition, and reflecting the history of domestication, the animals have changed, their reality quite different to that of their forebears 10,000 or more years ago. The painting also depicts the place of animals' lives in the city; humans and animals are interdependent. And in the increasingly global world, humans and animals are increasingly implicated in each other's futures in ways very few people understand anymore.

Humans live off the backs of farm animals; their productivity has provided our affluence, their compromises our benefits, whether as sources of food, commerce and so forth. And some of those compromises, such as the growth in productivity – increased stocking densities and efficiencies in converting forage into animal products, for example – have brought concern for animal welfare and social expectations (guidelines, regulations, assurances etc.). In other words, animal welfare is a socially constructed and driven issue – not just what the animal experiences or feels, but what society agrees on – regarding both what the animal experiences and, more importantly, whether any compromises to those experiences are justified by the wider benefits. The tensions in this construction are evident in the balance between caring and killing, the extent of humans' upper hand in the relationship. The sweet-spot between benefits to humans and the discomfort and suffering of animals, and any discomfort humans might experience, reverberates in responses to the 'noises' about what animal welfare society is prepared to accept, tolerate, negotiate or dramatize.

Associated with the intensification of farming of the last 10,000 years have, in part, been the 'demands' of society so that what the animal

Figure 7.1 *City with Animals (Stadt mit Tieren)* by Max Ernst (*Courtesy Solomon R. Guggenheim Museum, New York, Estate of Karl Nierendorf*). (Available in colour in the plate section between pages 178–179.)

experiences is the result of both the proximate (e.g. barren and confined environments) and the ultimate (e.g. the financial or economic pressures driving more intensive farming) effects of humans living off animals' backs. While animal welfare is socially driven, society's responses to poor animal welfare are, in the main, directed at the people in charge of animals, the proximate influences. The ultimate drivers remain largely unacknowledged, partly because most people are disconnected or estranged from farm animals, yet experience, rightly, emotional responses to instances of poor animal welfare. They lack knowledge of the context of human and animal interdependence. In the absence of that context, seeing, hearing, smelling and imagining are important, but the disconnection means they are viewed or understood in isolation, at least partially, from the benefits they receive from animals.

What history shows us is that farming, the human population and society's wealth and culture are intricately linked. It is the trade-offs that are the source of the tension. While it is always possible to improve animals' conditions, or quality of life, who should bear those costs? What the market, and any distortions like subsidies and failures to account for externalities, also do is allow us to 'forget' that link, so that consumers, of food and commerce especially, can be the elephant in the room leaving the animals to bear the costs. Paradoxically, the more farming has produced, the more people it has supported and the more affluent we have become, the more society determines what standards production must adhere to and the greater the need for monitoring and enforcement requiring greater resources from increased production! Like a pyramid or Ponzi scheme, are we borrowing from the future to pay for the present?

Another representation of the relationship is Banksy's *Trolley Hunters* (Figure 7.2). The hunter-gatherers, part of the natural world, can be seen as part of the ancient contract – a reliance on animals, but also a respect for them and the natural world, borne of seeing themselves as part of that world. One would not, however, want to live like a hunter-gatherer; it is important not to romanticize the life of a hunter-gatherer, for example, few possessions and a shorter life, ignoring the tensions, contradictions and complexities they too faced, and the 'truths' that may be more confusing than useful

Figure 7.2 *The Trolley Hunters (Courtesy Banksy)*. (Available in colour in the plate section between pages 178–179.)

(Preece, 1999). In contrast, the trolleys represent humans in the modern world, the affluent contract, their physical absence, apart from the symbols of consumption, the object of economic activity, reflecting the disconnection between humans and the natural world. Furthermore, this absence or disconnection means modern western humans can be devoid of social restraint and understanding borne of specialization, trading and transportation, and consumption driven not just by need, but also by 'our aspirations and hopes, our identities and secret cravings, our anxieties and boredom' (Roberts, 2014). The band of trolleys further represent modern consumers separated from the natural world (few of us know how to obtain our food as hunter-gatherers did, food coming from the supermarket). Animals need to be part of our lives so that we may make informed decisions about them. It highlights the importance of the many initiatives attempting to reconnect ourselves with the natural world, how we see ourselves in the world.

In reflecting on the relationship between animals and humans, perhaps there is room to examine mankind's place more critically. As a dog breeder stated in the novel *The Story of Edgar Sawtelle* (Wroblewski, 2008): 'In the end, to create better dogs, we will have to become better people.' What then is a good human, within the context of a fair relationship with animals? This is something society could explore in a disinterested manner, without having to feel threatened or defensive about current interests or perspectives. The relationship, and animal welfare, are everyone's problem and everyone needs to determine and accept their responsibilities. It is suggested that some of the starting points might include viewing animal welfare from the complexity and wickedness of the problem, determining the limits of the relationship, that it is too big, too complex, too wicked and too important to be left to science and farming, individuals and interest groups. What do we need to enable both farm animals and humans to flourish? Can the off-farm become a real or effective part of the farm again through the efforts of the animal welfare 'industry'? Should societal income be more directly linked to the success of farming: all pulling in our belts in a bad year, all enjoying the fruits of a good year?

How would we rewrite that domestic, ancient or affluent contract? Perhaps begin by acknowledging the complexity of animal welfare and involve people as citizens with the aim of identifying a modern hunter-gatherer, one that acts with the environment, including animals, in mind. Leaving relationships to be determined by economics is not a readily imaginable successful strategy. It is, in contrast, one that could see the future excesses of the most intensive systems, pig, poultry and dairy, justified because of the economic benefits to society. The tension between ethics and economics is too simplistic and it overrides the important things, like relationships, that make us human. Maybe our lifestyles will have to change if we are to better balance the tensions of being committed to inexpensive and plentiful food yet loyal to the more romantic images of farm settings, despite them being unable to produce efficiently and inexpensively. Or the tensions between agrarian and economic values, proximate and ultimate drivers. Maybe we all must live the changes required rather than telling others what to do. And having written the contract, what values and social institutions would need to be promoted and developed to support it, to make it easier for people to do the right thing by animals: treating them with the dignity of a good life and a quick and painless death. Perhaps trust is one of the more pressing values modern society should examine, especially the balance between the 'blind trust' based on power, status and authority, and 'earned trust' that requires authorities to earn it, and to continue to earn it. The former is more paternalistic, the latter more egalitarian in nature. Whom do, and should you trust, and why?

For all but the last 600 of the 250,000–300,000 generations since hominids first emerged, humans have been hunter-gatherers, arguably more egalitarian and connected with the natural world than most western populations are now. Throughout this time, we have built psychological mechanisms to disconnect ourselves, to deal with the tensions involved in having to both care for and kill animals. However, those mechanisms have broken down as science, especially, demonstrated the similarities between animals and humans. When Jane Goodall sent word to Louis Leakey that one of her chimpanzees, David Greybeard, had made and used tools, Leakey's response was: 'WE MUST REDEFINE TOOL STOP REDEFINE MAN STOP OR ACCEPT CHIMPANZEES AS HUMAN' (Gerber, 2017). It highlighted perhaps one of the more important barriers humans had erected between ourselves and animals. Another, more complex barrier is the disconnect experienced by much of the modern western world. We do not look animals in the eye anymore and perhaps do not fully appreciate our impacts on them. Although it is dangerous to view the ideals of one time from the position of another, as Preece (1999) cautions, rather than redefining animals as human, admitting them to some sort of human circle, perhaps there is value in examining our place in the natural world – to be part of the natural world, as hunter-gatherers may once have seen themselves. And it is not just the impacts on animals. Lawler (2016) tells of the villagers of Kirikongo in West Africa that did not follow the familiar pattern of elites developing in small villages and progressing through to political and religious leaders in cities and the emergence of empires. Rejecting farming cattle, typically the domain of men, they created a different society by farming chickens, making it easier for many people and especially women, not just the powerful elite, to participate in sacrifices. In other words, chickens, as they were relatively inexpensive and quickly

Figure 7.3 Bully Todd and a young bull whisperer (*Photo courtesy Fiona Fisher*). (Available in colour in the plate section between pages 178–179.)

reproduce, enabled a rich spiritual life within a more egalitarian society, the nature of which is still apparent in modern people living in the area.

The relationship humans have with farm animals (e.g. Figure 7.3) is special. While we cannot live without having some sort of impact on them, it is difficult to imagine what sort of a world it would be without farm animals. In accepting that, then empowering people to care for them, is imperative: providing them with the time, resources and confidence, especially younger generations, to develop the skills and empathy that enable and support the relationships society deems acceptable.

References

Aaltola, E. (2005) Animal ethics and interest conflicts. *Ethics & the Environment* 10, 19–48.
Adams, D. (1980) *The Ultimate Hitchhiker's Guide: The Restaurant at the End of the Universe*. Random House, New York, NY.
Adams, R. (1972) *Watership Down*. Penguin, Harmondsworth, UK.
Agger, J.F. and Willeberg, P. (1991) Production and mortality in dairy cows from 1960–1990: time series analysis of ecological data. In: *Proceedings of the 6th International Symposium of Veterinary Epidemiology and Economics*, Ottawa, Canada, pp. 357–360.
Aiken, W. (1991) The goals of agriculture. In: Blatz, C.V. (ed.) *Ethics and Agriculture. An Anthology on Current Issues in World Context*. University of Idaho Press, Moscow, ID, pp. 56–62.
Alexis, A. (2015) *Fifteen Dogs. A Novel*. Serpent's Tail, London, UK.
Alpigiani, I., Bacci, C., Keeling, L.J., Salman, M.D., Brindani, F., Pomgolini, S., Hitchens, P.L. and Bonardi, S. (2016) The associations between animal-based welfare measures and the presence of indicators of food safety in finishing pigs. *Animal Welfare* 25, 355–363.
Amory, J.R., Barker, Z.E., Wright, J.L., Mason, S.A., Blowey, R.W. and Green, L.E. (2008) Associations between sole ulcer, white line disease and digital dermatitis and the milk yield of 1824 dairy cows on 30 dairy cow farms in England and Wales from February 2003–November 2004. *Preventive Veterinary Medicine* 83, 381–391.
Amos, N. and Sullivan, R. (2016) The Business Benchmark on Farm Animal Welfare 2016 Annual Report. Available at: https://www.bbfaw.com/media/1450/bbfaw-2016-report.pdf (accessed 7 April 2018).
Anderson, E. (2004) Animal rights and the values of nonhuman life. In: Sunstein, C.R. and Nussbaum, M.C. (eds) *Animal Rights. Current Debates and New Directions*. Oxford University Press, Oxford, UK, pp. 277–298.
Anderson, W.T. (1990) Food without farms. The biotech revolution in agriculture. *The Futurist* 24, 16–22.
Andrade, S.B. and Anneberg, I. (2014) Farmers under pressure. Analysis of the

social conditions of cases of animal neglect. *Journal of Agricultural and Environmental Ethics* 27, 103–126.

Animal Welfare Institute (undated) A consumer's guide to food labels and animal welfare. https://awionline.org/content/consumers-guide-food-labels-and-animal-welfare (accessed 14 June 2018).

Animalia (2016) Funny cats scared of cucumbers - cat vs cucumber compilation. https://www.youtube.com/watch?v=JsysNml153M (accessed 13 June 2018).

Anneberg, I., Vaarst, M. and Sørensen, J.T. (2012) The experience of animal welfare inspections as perceived by Danish livestock farmers: a qualitative research approach. *Livestock Science* 147, 49–58.

Anonymous (1978) Slaughter protest on Invercargill streets 1978. Available at: http://www.invsoc.org.nz/slaughterprotest/ (accessed 7 April 2018).

Anonymous (1988) Readers respond to Catalin Valentin's lamb. *Utne Reader*, September/October, pp. 5–6.

Anonymous (2002a) Huge turnout for countryside march. *BBC News*, 22 September. Available at: http://news.bbc.co.uk/2/hi/uk_news/2274129.stm (accessed 7 April 2018).

Anonymous (2002b) Thousands march for countryside. *The Telegraph*, 22 September. Available at: https://www.telegraph.co.uk/news/1407980/Thousands-march-for-countryside.html (accessed 7 April 2018).

Anonymous (2005) Shearing swallows 35% of wool income. *Meat & Wool* 2 (2), 12.

Anonymous (2007) Snapshot of farming in the UK. *BBC*, 1 October. Available at: http://news.bbc.co.uk/2/hi/uk_news/magazine/6919829.stm (accessed 25 March 2018).

Anonymous (2012a) Production up, profit down: what's going on? *Straight Furrow*, 8 May, p. 23.

Anonymous (2012b) Sharemilking in 2012. Modern approach to industry's traditional path. *Inside Dairy*, May, pp. 2–7.

Anonymous (2014) Celebrating 50 years of the DairyNZ Economic Survey. *Inside Dairy*, May 2014, p. 6.

Anonymous (2016) BVA renews calls for mandatory CCTV in slaughterhouses. *The Veterinary Record*, 21 May, p. 517.

Anonymous (undated, a) Weighing up the economics of dairy farms. A briefing by the World Society for the Protection of Animals. http://media.wspa.org.uk/UKWEB/NIMC_PR_downloads/FOR_THE_PRESS.pdf (accessed 12 June 2018).

Anonymous (undated, b) High steaks: a humane and sustainable 'farm to fork' beef system in the US. World Society for the Protection of Animals, 4 pp.

Anonymous (undated, c) Sheep-n-show. www.sheepnshow.co.nz/ (accessed 25 March 2018).

Anthony, R. (2003) The ethical implications of the human–animal bond on the farm. *Animal Welfare* 12, 505–512.

Anthony, R. (2009) Farming animals and the capabilities approach: understanding roles and responsibilities through narrative ethics. *Society and Animals* 17, 257–278.

Appleby, M.C. (1999) *What Should We Do about Animal Welfare?* Blackwell, Oxford, UK.

Appleby, M.C. and Hughes, B.O. (2005) *Animal Welfare.* CABI Publishing, Wallingford, UK.

Arey, D. and Brooke, P. (2006) *Animal Welfare Aspects of Good Agricultural Practice: Pig Production.* Compassion in World Farming Trust, Petersfield, UK.

Arman, P. (1974) A note on parturition and maternal behaviour in captive red deer (Cervus elaphus L.). *Journal of Reproduction and Fertility* 37, 87–90.

Arman, P., Hamilton, W.J. and Sharman, G.A.M. (1978) Observations on the calving of free-ranging tame red deer (*Cervus elaphus*). *Journal of Reproduction and Fertility* 54, 279–283.

Armstrong, J. (2009) *In Search of Civilisation. Remaking a Tarnished Idea.* Allen Lane, London, UK.

Armstrong, K. (2009) *The Case for God. What Religion Really Means.* Vintage, London, UK.

Ascione, F.R., Weber, C.V., Thompson, T.M., Heath, J., Maruyama, M. and Hayashi, K. (2007) Battered pets and domestic violence. Animal abuse reported by women experiencing intimate violence and non-abused women. *Violence Against Women* 13, 354–373.

Atwood, M. (2008) *Payback. Debt and the Shadow Side of Wealth.* Bloomsbury, London, UK.

Autio, M., Autio, J., Kuismin, A., Ramsingh, B., Kylkilahti, E. and Valros, A. (2018) Bringing farm animal welfare to the consumer's plate: the quest for food business to enhance transparency, labelling and consumer education. In: Amos, N. and Sullivan, R. (eds) *The Business of Farm Animal Welfare.* Greenleaf Publishing, Routledge, London, UK, pp. 120–136.

Bailey, M. (1988) The rabbit and the medieval East Anglian economy. *Agricultural History Review* 36, 1–20.

Balon, E.K. (2004) About the oldest domesticates among fishes. *Journal of Fish Biology* 65, 1–27.

Banner, M., Bulfield, G., Clark, S., Gormally, L., Hignett, P., Kimbell, H., Milburn, C. and Moffitt, J. (1995) *Report of the Committee to Consider the Ethical Implications of Emerging Technologies in the Breeding of Farm Animals.* HMSO, London, UK.

Baran, B.E., Allen, J.A., Rogelberg, S.G., Spitzmüller, C., DiGiacomo, N.A., Webb, J.B., Carter, N.T., Clark, O.L., Teeter, L.A. and Walker, A.G. (2009) Euthanasia-related strain and coping strategies in animal shelter employees. *Journal of the American Veterinary Medical Association* 235, 83–88.

Barker, G. (2006) *The Agricultural Revolution in Prehistory. Why Did Foragers Become Farmers?* Oxford University Press, Oxford, UK.

Baxter, E.M., Rutherford, K.M.D., D'Eath, R.B., Arnott, G., Turner, S.P., Sandøe, P., Moustsen, V.A., Thorup, F., Edwards, S.A. and Lawrence, A.B. (2013) The welfare implications of large litter size in the domestic pig II: management factors. *Animal Welfare* 22, 219–238.

Beauchamp, T.L. and Childress, J.F. (1994) *Principles of Biomedical Ethics*, 4th edn. Oxford University Press, New York, NY.

Beckett, J.V. (1989) *A History of Laxton: England's Last Open-Field Village*. Basil Blackwell, Oxford, UK.

Beckett, J.V. (1990) *The Agricultural Revolution*. Basil Blackwell, Oxford, UK.

Beeton, I. (1960) *Mrs Beeton's Cookery and Household Management*, 11th impression. Ward Lock, London, UK.

Belfer-Cohen, A. (1991) The Natufian in the Levant. *Annual Review of Anthropology* 20, 167–186.

Bell, E., Norwood, F.B. and Lusk, J.L. (2017) Are consumers wilfully ignorant about animal welfare? *Animal Welfare* 26, 399–402.

Belyaev, D.K. and Trut, L.N. (1975) Some genetic and endocrine effects of selection for domestication in silver foxes. In: Fox, M.W. (ed.) *The Wild Canids. Their Systematics, Behavioral Ecology and Evolution*. Van Nostrand Reinhold, New York, NY, pp. 416–426.

Bennett, A.F., Huey, R.B., John-Alder, H. and Nagy, K.A. (1984) The parasol tail and thermoregulatory behavior of the Cape ground squirrel xerus inauius. *Physiological Zoology* 57, 57–62.

Bennett, R.M. (1997) Economics. In: Appleby, M.C. and Hughes, B.O. (eds) *Animal Welfare*. CABI Publishing, Wallingford, UK, pp. 235–248.

Benson, R. (2005) *The Farm. The Story of One Family and the English Countryside*. Penguin, London, UK.

Benton, T. (1993) *Natural Relations. Ecology, Animal Rights and Social Justice*. Verso, London, UK.

Berger, J. (1980) *About Looking*. Vintage International, New York, NY.

Berry, W. (1984) Whose head is using the farmer and whose head is the farmer using? In: Jackson, W., Berry, W. and Colman, B. (eds) *Meeting the Expectations of the Land. Essays in Sustainable Agriculture and Stewardship*. North Point Press, San Francisco, CA, pp. 19–30.

Berry, W. (1996) *The Unsettling of America and Agriculture*, 3rd edn. Sierra Club, San Francisco, CA.

Beveridge, Sir W. (1942) *Social Insurance and Allied Services*. Her Majesty's Stationery Office, London, UK. Available at: https://www.ncbi.nlm.nih.gov/pmc/articles/PMC2560775/pdf/10916922.pdf (accessed 25 March 2018).

BeVier, G.W. and Lautner, B. (1994) Realities of contemporary livestock production. *Journal of the American Veterinary Medical Association* 204, 369–371.

Bezant, L. (1999) In praise of the wild boar. *Journal of the Royal Agricultural Society of England* 160, 90–98.

Bielman, J. (2005) Technological innovation in Dutch cattle breeding and dairy farming, 1850–2000. *The Agricultural History Review* 53, 229–250.

Blaxter, K.L., Kay, R.N.B., Sharman, G.A.M., Cunningham, J.M.M. and Hamilton, W.J. (1974) *Farming the Red Deer. The First Report of an Investigation by the Rowett Research Institute and the Hill Farming Research Organisation*. HMSO, Edinburgh, UK.

Blokhuis, H., Miele, M., Veissier, I. and Jones, B. (2013) *Improving Farm Animal Welfare. Science and Society Working Together: The Welfare Quality Approach*. Wageningen Academic, Wageningen, the Netherlands.

Body, R. (1984) *Farming in the Clouds*. Temple Smith, London, UK.
Body, R. (1991) *Our Food, Our Land. Why Contemporary Farming Practices Must Change*. Rider, London, UK.
Bohm, D. (1996) *On Dialogue*. Routledge, London, UK.
Boivin, X., Lensink, J., Tallet, C. and Veissier, I. (2003) Stockmanship and farm animal welfare. *Animal Welfare* 12, 479–492.
Bökönyi, S. (1974) *History of Domestic Mammals in Central and Eastern Europe*. Adadémiai Kiadó, Budapest, Hungary.
Boland, M.J., Rae, A.N., Vereijken, J.M., Meuwissen, M.P.M., Fischer, A.R.H., van Boekel, M.A.J.S., Rutherfurd, S.M., Gruppen, H., Moughan, P.J. and Hemdriks, W.M. (2013) The future supply of animal-derived protein for human consumption. *Trends in Food Science & Technology* 29, 62–73.
Boogaard, B.K., Boekhorst, L.J.S., Oosting, S.J. and Sørensen, J.T. (2011) Socio-cultural sustainability of pig production: citizen perceptions in the Netherlands and Denmark. *Livestock Science* 140, 189–200.
Botreau, R., Winckler, C., Verlarde, A., Butterworth, A., Dalmau, A., Keeling, L. and Veissier, I. (2013) Integration of data collected on farms or at slaughter to generate an overall assessment of animal welfare. In: Blokhuis, H., Miele, M., Veissier, I. and Jones, B. (eds) *Improving Farm Animal Welfare. Science and Society Working Together: The Welfare Quality Approach*. Wageningen Academic, Wageningen, the Netherlands, pp. 147–174.
Boyd, B. (2009) *On the Origin of Stories. Evolution, Cognition, and Fiction*. Belknap Press, Cambridge, MA.
Boztas, S. (2016) Do cows get seasick? Welcome to Rotterdam's floating dairy farm. *The Guardian*, 4 July. Available at: https://www.theguardian.com/sustainable-business/2016/jul/04/do-cows-get-seasick-rotterdam-floating-dairy-farm-netherlands (accessed 6 March 2018).
Brache, M.B.M. and Hopster, H. (2006) Assessing the importance of natural behaviour for animal welfare. *Journal of Agricultural and Environmental Ethics* 19, 77–89.
Bradshaw, J.W.S. and Paul, E.S. (2010) Could empathy for animals have been an adaptation in the evolution of Homo sapiens? *Animal Welfare* 19(S), 107–112.
Brambell, F.W.R. (1965) *Report of the Technical Committee to Enquire into the Welfare of Animals Kept under Intensive Livestock Husbandry Systems*. Her Majesty's Stationery Office, London, UK.
Bray, A.R. and Gonzalez-Macuer, E. (2010) New Zealand sheep and wool industries. In: Cottle, D.J. (ed.) *International Sheep and Wool Handbook*. Nottingham University Press, Nottingham, UK, pp. 73–83.
Bronowski, J. (1973) *The Ascent of Man*. BBC, London, UK.
Broom D. (2003) *The Evolution of Morality and Religion*. Cambridge: Cambridge University Press, Cambridge, UK.
Broom, D.M. (2011) A history of animal welfare science. *Acta Biotheoritica* 59, 121–137.
Broom, D.M. and Fraser, A.F. (2015) *Domestic Animal Behaviour and Welfare*, 5th edn. CABI, Wallingford, UK.

Broom, D.M. and Johnson, K.G. (1993) *Stress and Animal Welfare*. Chapman & Hall, London, UK.

Brophy, B. (1966) *Don't Never Forget. Collected Views and Reviews*. Jonathan Cape, London, UK.

Brophy, B. (1971) In pursuit of a fantasy. In: Godlovitch, S., Godlovitch, R. and Harris, J. (eds) *Animals, Men and Morals. An Enquiry into the Maltreatment of Non-Humans*. Victor Gollancz, London, UK, pp. 125–145.

Browne, W.P., Skees, J.R., Swanson, L.E., Thompson, P.B. and Unnevehr, L.J. (1992) *Sacred Cows and Hot Potatoes. Agrarian Myths in Agricultural Policy*. Westview, Boulder, CO.

Bull, S.A., Thomas, A., Humphrey, T.A., Ellis-Iversen, J., Cook, A.J., Lovell, R. and Jorgensen, F. (2008) Flock health indicators and Campylobacter spp in commercial housed broilers reared in Great Britain. *Applied Environmental Microbiology* 74, 5408–5413.

Burkhardt, C.A., Cherry, J.A., Van Krey, H.P. and Siegel, P.B. (1983) Genetic selection for growth rate alters hypothalamic satiety mechanisms in chickens. *Behaviour Genetics* 13, 295–300.

Bustad, L.K. (1988) Living together: people, animals, environment—a personal historical perspective. *Perspectives in Biology and Medicine* 31, 171–184.

Campbell, J. (1968) *The Hero with a Thousand Faces*. Princeton University Press, Princeton, NJ.

Campbell, J. (1984) *The Way of the Animal Powers. Historical Atlas of World Mythology*. Times Books, London, UK.

Capps, O. Jr. and Park, J.L. (2003) Food retailing and food service. *Veterinary Clinics of North America: Food Animal Practice* 19, 445–461.

Capra, F. (2002) *The Hidden Connections. A Science for Sustainable Living*. Harper Collins, London, UK.

Carruthers, P. (1992) *The Animals Issue. Moral Theory in Practice*. Cambridge University Press, Cambridge, UK.

Carson, R. (1962) *Silent Spring*. Houghton Mifflin, Boston, MA.

Caughley, G. (1983) *The Deer Wars. The Story of Deer in New Zealand*. Heinemann, Auckland, NZ.

Caughley, G. (1988) Control of wild animals. In: Newton, A.E. (ed.) *The Future of New Zealand's Wild Animals. Seminar 2000*. New Zealand Deerstalkers Association, Wellington, NZ, pp. 101–103.

Chagas, L.M., Bass, J.J., Blache, D., Burke, C.R., Kay, J.K., Lindsay, D.R., Lucy, M.C., Martin, G.B., Meier, S., Rhodes, F.M., Roche, J.R., Thatcher, W.W. and Webb, R. (2007) New perspectives on the roles of nutrition and metabolic profiles in the subfertility of high producing dairy cows. *Journal of Dairy Science* 90, 4022–4032.

Challies, C.N. (1985) Establishment, control, and commercial exploitation of wild deer in New Zealand. In: Fennessy, P.F. and Drew, K.R. (eds) *Biology of Deer Production*. Royal Society of New Zealand, Wellington, NZ, pp. 23–36.

Chambers, T. (1999) *The Fiction of Bioethics. Cases as Literary Texts*. Routledge, New York, NY.

Cheeke, P.R. (1999) Shrinking membership in the American Society of Animal Science: does the discipline of poultry science give us some clues? *Journal of Animal Science* 77, 2031–2038.

Cherfas, J. (2012) Heart-stopping coincidence. Available at: http://agro.biodiver.se/2012/12/heart-stopping-coincidence/ (accessed 25 March 2018).

Chriel, M. and Dietz, H.H. (2003) Medication of production animals – cure of malfunctioning animals or production systems? *Acta Veterinaria Scandinavica, Supplementum* 98, 65–70.

Christensen, T., Lawrence, A., Lund, M., Stott, A. and Sandøe, P. (2012) How can economists help improve animal welfare? *Animal Welfare* 21 (Supplement 1), 1–10.

Christiansen, S.B. and Sandøe, P. (2000) Bioethics: limits to the interference with life. *Animal Reproduction Science* 60–61, 15–19.

Clark, B., Stewart, G.B., Panzone, L.A., Kyriazakis, I. and Frewer, L.J. (2017) Citizens, consumers and farm animal welfare: a meta-analysis of willingness-to-pay studies. *Food Policy* 68, 112–127.

Clark, D.A., Caradus, J.R., Monaghan, R.M., Sharp, P. and Thorrold, B.S. (2007) Issues and options for future dairy farming in New Zealand. *New Zealand Journal of Agricultural Research* 50, 203–221.

Clark, S.R.L. (1998) Making up animals: the view from science fiction. In: Holland. A. and Johnson, A. (eds) *Animal Biotechnology and Ethics*. Chapman & Hall, London, UK, pp. 209–224.

Clark, S.R.L. (1999) *The Political Animal. Biology, Ethics and Politics*. Routledge, London, UK.

Clements, J. (2015) How science fiction helps us reimagine our moral relations with animals. *Journal of Animal Ethics* 5, 181–187.

Clutton-Brock, J. (1992) How the wild beasts were tamed. *New Scientist*, 15 February, pp. 31–33.

Clutton-Brock, J. (1999) *A Natural History of Domesticated Mammals*. Cambridge University Press, Cambridge, UK.

Cochran, G. and Harpending, H. (2009) *The 10,000 Year Explosion. How Civilisation Accelerated Human Evolution*. Basic Books, New York, NY.

Coetzee, J.M. (2000) *Disgrace*. Vintage, London, UK.

Collias, N.E. and Collias, E.C. (1967) A field study of the red jungle fowl in north-central India. *The Condor* 69, 360–386.

Columella, L.J.M. (undated) *On Agriculture*. Available at: https://archive.org/stream/onagriculturewit02coluuoft/onagriculturewit02coluuoft_djvu.txt (accessed 25 March 2018).

Compton, C.W.R., Heuer, C., Thomsen, P.T., Carpenter, T.E., Phyn, C.V.C. and McDougall, S. (2017) A systematic literature review and meta-analysis of mortality and culling in dairy cattle. *Journal of Dairy Science* 100, 1–16.

Comstock, G. (1987) *Is There a Moral Obligation to Save the Family Farm?* Iowa State University Press, Ames, IA.

Comstock, G.L. (2000) An alternative ethic for animals. In: Hodges, J. and Han, I.K. (eds) *Livestock, Ethics and Quality of Life*. CAB International, Wallingford, UK, pp. 99–118.

Convery, I., Bailey, C., Mort, M. and Baxter, J. (2005) Death in the wrong place? Emotional geographies of the UK 2001 foot and mouth disease epidemic. *Journal of Rural Studies* 21, 99–109.

Cooper, M.R., Barton, G.T. and Brodell, A.P. (1947) *Progress of Farm Mechanization.* USDA, Washington, D.C.

Cooper, O. (2006) The more it changes ... *Farmers Weekly*, 15 December, p. 32.

Coppinger, R.P. and Smith, C.K. (1983) The domestication of evolution. *Environmental Conservation* 10, 283–292.

Cornish, A., Raubenheimer, D. and McGreevy, P. (2016) What we know about the public's level of concern for farm animal welfare in food production in developed countries. *Animals* 6, 74.

Cottingham, J. (1978) 'A brute to the brutes?': Descartes' treatment of animals. *Philosophy* 53, 551–559.

Covey, S.R. (1989) *The Seven Habits of Highly Effective People. Restoring the Character Ethic.* Business Library, Melbourne, Australia.

Cowie, G., Moore, G.H., Fisher, M.W. and Taylor, M. (1985) Calving behaviour in farmed red deer. *Proceedings of the Deer Branch New Zealand Veterinary Association* 2, 143–154.

Crabtree, P.J. (1993) Early animal domestication in the Middle East and Europe. *Archaeological Method and Theory* 5, 201–245.

Crump, B. (1960) *A Good Keen Man.* A.H. & A.W. Reed, Wellington, NZ.

Curry, J.M. (2002) Care theory and 'caring' systems of agriculture. *Agriculture and Human Values* 19, 119–131.

Dale, T. and Carter, V.G. (1955) *Topsoil and Civilization.* University of Oklahoma Press, Norman, OK.

Darwash, A.O. and Lamming, G.E. (1997) Abnormal ovarian patterns as a cause of subfertility in dairy cows: protocols for early detection and treatment. *Cattle Practice* 5, 3–7.

Darwin, C. (1859) *On the Origin of the Species.* Oxford University Press edition (2008), Oxford, UK.

Davis, W. (2009) *The Wayfinders. Why Ancient Wisdom Matters in the Modern World.* Anansi, Toronto, Canada.

Dawkins, M.S. (1980) *Animal Suffering. The Science of Animal Welfare.* Chapman & Hall, London, UK.

Dawkins, M.S. (2008) The science of animal suffering. *Ethology* 114, 937–945.

Dawkins, M.S. (2016) Animal welfare with and without consciousness. *Journal of Zoology* 301, 1–10.

de Bakker, E. and Dagevos, H. (2012) Reducing meat consumption in today's consumer society: questioning the citizen-consumer gap. *Journal of Agricultural and Environmental Ethics* 25, 877–894.

de Boer, I.J.M., Brom, F.W.A. and Vorstenbosch, J.M.G. (1995) An ethical evaluation of animal biotechnology: the case of using clones in dairy cattle breeding. *Animal Science* 61, 453–463.

DeBoer, S.P., Garner, J.P., McCain, R.R., Lay, D.C. Jr, Eicher, S.D. and Marchant-Ford, J.N. (2015) An initial investigation into the effects of isolation and enrichment on the welfare of laboratory pigs housed in the PigTurn® system, assessed using tear staining, behaviour, physiology and haematology. *Animal Welfare* 24, 15–27.

Derr, M. (2012) *How the Dog Became the Dog. From Wolves to Our Best Friends.* Scribe, Melbourne, Australia.

Descartes, R. (1976) Animals are machines. In: Regan, T. and Singer, P. (eds) *Animal Rights and Human Obligations.* Prentice Hall, Englewood Cliffs, NJ, pp. 60–66.

Devitt, C., Kelly, P., Blake, M., Hanlon, A. and More, S.J. (2014) An investigation into the welfare element of on-farm animal welfare incidents in Ireland. *Sociologia Ruralis* 55, 400–416.

Dewey, P. (1989) Agriculture. In: Pope, R. (ed.) *Atlas of British Social and Economic History since c. 1700.* Routledge, London, UK, pp. 1–22.

Diamond, J. (1987) The worst mistake in the history of the human race. *Discover,* May, pp. 64–66.

Diamond, J. (1991) *The Rise and Fall of the Third Chimpanzee.* Vintage, London, UK.

Diamond, J. (1997) *Guns, Germs and Steel. The Fates of Human Societies.* Jonathan Cape, London, UK.

Diamond, J. (2002) Evolution, consequences and future of plant and animal domestication. *Nature* 418, 700–706.

Dick, P.K. (1968) *Do Androids Dream of Electric Sheep?* Weidenfeld & Nicolson, London, UK.

Dillard, J. (2008) A slaughterhouse nightmare: psychological harm suffered by slaughterhouse employees and the possibility of redress through legal reform. *Georgetown Journal on Poverty Law & Policy* 15, 391–409.

Dobson, H., Smith, R.F., Royal, M.D., Knight, C.H. and Sheldon, I.M. (2007) The high-producing dairy cow and its reproductive performance. *Reproduction in Domestic Animals* 42 (Suppl. 2), 17–23.

Dobson, H., Walker, S.L., Morris, M.J., Routly, J.E. and Smith, R.F. (2008) Why is it getting more difficult to successfully artificially inseminate dairy cows? *Animal* 2, 1104–1111.

Donham, K.J. (1998) The impact of industrial swine production on human health. In: Thu, K.M. and Durrenberger, E.P. (eds) *Pigs, Profits, and Rural Communities.* State University of New York Press, Albany, NY, pp. 73–83.

Donham, K.J. and Thu, K.M. (1993) Relationships of agricultural and economic policy to the health of farm families, livestock, and the environment. *Journal of the American Veterinary Medical Association* 202, 1084–1091.

Donnelley, S. (1995) The art of moral ecology. *Ecosystem Health* 1, 170–176.

Donnelly, C.A., Woodroffe, R., Cox, D.R., Bourne, F.J., Cheeseman, C.L., Clifton-Hadley, R.S., Wei, G., Gettinby, G., Gilks, P., Jenkins, H., Johnston, W.T., Le Fevre, A.M., McInerney, J.P. and Morrison, W.I. (2006) Positive and negative effects of widespread badger culling on tuberculosis in cattle. *Nature* 439, 843–846.

Dosa, D.M. (2007) A day in the life of Oscar the cat. *New England Journal of Medicine* 357, 328–329.

Doughty, A.K., Coleman, G.J., Hinch, G.N. and Doyle, R.E. (2017) Stakeholder

perceptions of welfare issues and indicators for extensively managed sheep in Australia. *Animals* 7, 28.
Downs, J. (1960) Domestication: an examination of the changing social relationships between man and animals. *Kroeber Anthropological Society* 22, 18–67.
Doyon, M., Bergeran, S., Cranfield, J., Tamini, L. and Criner, G. (2016) Consumer preferences for improved hen housing: is a cage a cage? *Canadian Journal of Agricultural Economics* 64, 739–751.
Drabenstott, M. (1995) Agricultural industrialization: implications for economic development and public policy. *Journal of Agricultural and Applied Economics* 27, 13–20.
Druett, J. (1983) *Exotic Intruders. The Introduction of Plants and Animals into New Zealand*. Heinemann, Auckland, NZ.
Duncan, I.J.H., Savory, C.J. and Wood-Gush, D.M. (1978) Observations on the reproductive behaviour of domestic fowl in the wild. *Applied Animal Ethology* 4, 29–42.
Duncan, I.J.H. and Wood-Gush, D.G.M. (1971) Frustration and aggression in the domestic fowl. *Animal Behaviour* 19, 500–504.
Durning, A.T. and Brough, H.B. (1995) Animal farming and the environment. In: Cooper, D.E. and Palmer, J.A. (eds) *Just Environments. Intergenerational, International and Interspecies Issues*. Routledge, London, UK, pp. 149–164.
Dwyer, C.M. (2008) *The Welfare of Sheep*. Springer, Dordrecht, the Netherlands.
Dwyer, C.M. and Lawrence, A.B. (2005) A review of the behavioural and physiological adaptations of hill and lowland breeds of sheep that favour lamb survival. *Applied Animal Behaviour Science* 92, 235–260.
Early, R. (1998) Farm assurance – benefit or burden? *Journal of the Royal Agricultural Society of England* 159, 32–43.
Ellis, A.B. (1894) *Yoruba-Speaking Peoples of the Slave Coast of West Africa*. Available at: http://www.obafemio.com/uploads/5/1/4/2/5142021/yoruba_speaking_peoples_of_west_africa.pdf (accessed 4 April 2018).
Espmark, Y. and Langvatn, R. (1979) Cardiac responses in alarmed red deer calves. *Behavioural Processes* 4, 179–186.
Espmark, Y. and Langvatn, R. (1985) Development and habituation of cardiac and behavioral responses in young red deer calves *Cervus elaphus* exposed to alarm stimuli. *Journal of Mammalogy* 66, 702–711.
Esslemont, R.J. and Kossaibati, M.A. (1996) Incidence of production diseases and other health problems in a group of dairy herds in England. *The Veterinary Record* 139, 486–490.
Fagan, B. (2015) *The Intimate Bond. How Animals Shaped Human History*. Bloomsbury, New York, NY.
Fairweather, J.R. and Hunt, L.M. (2011) Can farmers map their farm system? Causal mapping and the sustainability of sheep/beef farms in New Zealand. *Agriculture and Human Values* 28, 55–66.
Farm Animal Welfare Council (2009) *Opinion on the Welfare of the Dairy Cow*. FAWC, London, UK.

Farm Animal Welfare Committee (2011) *Education, Communication and Knowledge Application in Relation to Farm Animal Welfare.* FAWC, London, UK.

Farm Animal Welfare Committee (2014) *Evidence and the Welfare of Farmed Animals. Part 1: The Evidence Base.* FAWC, London, UK.

Farm Animal Welfare Committee (2015) *Opinion on CCTV in Slaughterhouses.* FAWC, London, UK.

Farm Animal Welfare Committee (2016) *Opinion on the Links between the Health and Wellbeing of Farmers and Farm Animal Welfare.* FAWC, London, UK.

Ferguson, N. (2011) *Civilization. The Six Killer Apps of Western Power.* Penguin, London, UK.

Ferguson, N. (2012) *The Great Degeneration. How Institutions Decay and Economies Die.* Allen Lane, London, UK.

Finlayson, C. (2009) *The Humans Who Went Extinct. Why Neanderthals Died out and We Survived.* Oxford University Press, Oxford, UK.

Finney, T. (2007) A slaughterhouse. *Food Ethics* 2 (4), 15.

Firbank, L.G., Elliot, J., Drake, B., Cao, Y. and Gooday, R. (2013) Evidence of sustainable intensification among British farms. *Agriculture, Ecosystem and Environment* 173, 58–65.

Fisher, M. (1961) The postman. *New Zealand Department of Education Correspondence School Annual Magazine* 34, p. 22.

Fisher, M.W. (2002) Skeletons and sovereigns in the cupboard – learning from our myths. In: Fisher, M., Marbrook, J. and Sutherland, G. (eds) *Learning, Animals and the Environment: Changing the Face of the Future.* ANZCCART, Wellington, NZ, pp. 50–53.

Fisher, M.W. (2007) Shelter and welfare of pastoral animals in New Zealand. *New Zealand Journal of Agricultural Research* 50, 347–359.

Fisher, M.W. (2009) Defining animal welfare – does consistency matter? *New Zealand Veterinary Journal* 57, 71–73.

Fisher, M. (2010) A method for considering the acceptability of novel biotechnologies for the control of brushtail possums. *Kōtuitui: New Zealand Journal of Social Sciences* 5, 41–52.

Fisher, M.W. (2011) The welfare of extensively farmed livestock. *Proceedings of the New Zealand Society of Animal Production* 71, 181–185.

Fisher, M., Bloomer, D., Wedderburn, L. and Botha, N. (2009) *Cropping Farmer Perspectives of Maintaining and Enhancing Soil Quality on the Heretaunga Plains.* AgMARDT, Wellington, NZ.

Fisher, M.W. and Bryant, L.D. (1993) What might be the consequences of adapting wild animals, such as wapiti, to a farm environment? *Proceedings of the New Zealand Society of Animal Production* 53, 457–460.

Fisher, M., Diesch, T. and Orange, M. (2014) Scapegoats and tourists – science as the source of, and the answer to, ethical dilemmas. In: Sutherland, G. and Cragg, P. (eds) *Mixing it up – Ethics, Science and Adventure Tourism.* ANZCCART, Wellington, NZ, pp. 1–7.

Fisher, M.W. and Mellor, D.J. (2008) Developing a systematic strategy incorporating ethical, animal welfare and practical principles to guide the genetic improvement of dairy cattle. *New Zealand Veterinary Journal* 56, 100–106.

Fisher, M.W., Muir, P.D., Gregory, N.G., Thomson, B.C., Smith, N.B., Johnstone, P.D. and Bicknell, N. (2015) The effects of depriving feed to facilitate transport and slaughter in sheep – a case study of cull ewes held off pasture for different periods. *New Zealand Veterinary Journal* 63, 260–264.

Fisher, M.W., Wild, R.J., O'Grady, K.R., Guigou, M.J., Diesch, T.J., Jamieson, J.V., Ward, L.J. and Cross, N.J. (2017) The welfare of bobby calves sent for slaughter: a synopsis of the science literature within the context of New Zealand's export system and the problems and gaps in it. *Proceedings of the New Zealand Society of Animal Production* 77, 149–153.

Fisk, R. (2008) *The Age of the Warrior*. Harper Collins, London, UK.

Flannery, T.F. (1994) *The Future Eaters. An Ecological History of the Australasian Peoples*. Reed Books, Melbourne, Australia.

Flynn, C.P. (2001) Acknowledging the 'zoological connection': a sociological analysis of animal cruelty. *Society & Animals* 9, 71–87.

Flynn, J. (2012) *Fate & Philosophy. A Journey through Life's Great Questions*. Awa Press, Wellington, NZ.

Food Ethics Council (2008) Ethical labels and standards. Does differentiation drive progress? Available at: https://www.foodethicscouncil.org/uploads/publications/businessforum180308.pdf (accessed 7 April 2018).

Food Ethics Council and Pickett, H. (2014) *Farm Animal Welfare: Past, Present and Future. A Review of Farm Animal Welfare in the UK*. Freedom Food and RSPCA, London, UK.

Forero, J. (2009) Day of the gaucho waning in Argentina. *Washington Post*, 10 September. Available at: http://www.washingtonpost.com/wp-dyn/content/article/2009/09/09/AR2009090903211.html (accessed 7 April 2018).

Forsberg, E.-M. (2011) Inspiring respect for animals through the law? Current development in Norwegian animal welfare legislation. *Journal of Agricultural and Environmental Ethics* 224, 351–366.

Fox, M.W. (1986) *Agricide. The Hidden Farm and Food Crisis That Affects Us All*. Schocken, New York, NY.

Fox, M.W. (1992) *Superpigs and Wondercorn. The Brave New World of Biotechnology and Where It All May Lead*. Lyons & Burford, New York, NY.

Franklin, S. (2007) *Dolly Mixtures. The Remaking of Genealogy*. Duke University Press, Durham, NC.

Fraser, C.E., Smith, K.B., Judd, F., Humphreys, J.S. and Fragar, L.J. (2005) Farming and mental health problems and mental illness. *International Journal of Social Psychiatry* 51, 340–349.

Fraser, D. (1978) Observations on the behavioural development of suckling and early-weaned piglets during the first six weeks after birth. *Animal Behaviour* 26, 22–30.

Fraser, D. (1993) Assessing animal well-being: common sense, uncommon science. In: *Food Animal Well-being 1993 – Conference Proceedings and Deliberations*. USDA and Purdue University, West Lafayette, IN, pp. 37–54.

Fraser, D. (1995) Science, values and animal welfare: exploring the 'inextricable connection'. *Animal Welfare* 4, 103–117.

Fraser, D. (1999) Animal ethics and animal welfare science: bridging the two cultures. *Applied Animal Behaviour Science* 65, 171–189.

Fraser, D. (2001) The 'new perception' of animal agriculture: legless cows, featherless chickens, and a need for genuine analysis. *Journal of Animal Science* 79, 634–641.

Fraser, D. (2003) Assessing animal welfare at the farm and group level: the interplay of science and values. *Animal Welfare* 12, 433–443.

Fraser, D. (2005) *Animal Welfare and the Intensification of Animal Production. An Alternative Interpretation*. FAO, Rome, Italy.

Fraser, D. (2008) *Understanding Animal Welfare. The Science in Its Cultural Context*. Wiley-Blackwell, Oxford, UK.

Fraser, D. (2014) Could animal production become a profession? *Livestock Science* 169, 155–162.

Fraser, D., Duncan, I.J., Edwards, S.A., Grandin, T., Gregory, N.G., Guyonnet, V., Hemsworth, P.H., Huertas, S.M., Huzzey, J.M., Mellor, D.J., Mench, J.A., Marek Špinka, M. and Whay, H.R. (2013) General principles for the welfare of animals in production systems: the underlying science and its application. *The Veterinary Journal* 198, 19–27.

Fraser, D. and MacRae, A.M. (2011) Four types of activities that affect animals: implications for animal welfare science and animal ethics philosophy. *Animal Welfare* 20, 581–590.

Friedman, T.L. (1999) *The Lexus and the Olive Tree*. Harper Collins, London, UK.

Fukuoka, M. (1978) *The One-Straw Revolution: An Introduction to Natural Farming*. Rodale Press, New York, NY.

Fuseini, A., Wotton, S.B., Hadley, P.J. and Knowles, T.G. (2017) The compatibility of modern slaughter techniques with halal slaughter: a review of aspects of 'modern' slaughter methods that divide scholarly opinion within the Muslim community. *Animal Welfare* 26, 301–310.

Gaarder, J. (2015) *The World According to Anna*. Weidenfeld & Nicolson, London, UK.

Gallagher, L.M., Kliem, C., Beautrais, A.L. and Stallones, L. (2007) Suicide and occupation in New Zealand, 2001–2005. *International Journal of Occupational and Environmental Health* 14, 45–50.

Garner, R. (1993) *Animals, Politics and Morality*. Manchester University Press, Manchester, UK.

Gatward, G.J. (2001) *Livestock Ethics. Respect, and Our Duty of Care for Farm Animals*. Chalcombe, Lincoln, UK.

Gerber, T. (2017) Becoming Jane. *National Geographic* 232 (4), 30–51.

Gjerris, M. (2015) Willed blindness: a discussion of our moral shortcomings in relation to animals. *Journal of Agricultural and Environmental Ethics* 28, 517–532.

Glover, D. (1964) *Enter without Knocking: Selected Poems*. Pegasus Press, Christchurch, NZ.

Gocsik, E., Lansink, A.O., Voermans, G. and Saatkamp, H.W. (2015) Economic feasibility of animal welfare improvements in Dutch intensive livestock production: a comparison between broiler, laying hen, and fattening sectors. *Livestock Science* 182, 38–53.

Goldschmidt, W. (1947) *As You Sow. Three Studies in the Social Consequences of Agribusiness*. Allanheld Osum, Montclair, NJ.

Goldschmidt, W. (1998) The urbanization of rural America. In: Thu, K.M. and Durrenberger, E.P. (eds) *Pigs, Profits, and Rural Communities*. State University of New York Press, Albany, NY, pp. 183–198.

Gonzalez, P.R. (2016) Beef-mad Argentina preparing for unthinkable as meat costs soar. *Bloomberg*, 10 March. Available at: https://www.bloomberg.com/news/articles/2016-03-10/beef-mad-argentina-preparing-for-unthinkable-as-meat-costs-soar (accessed 7 April 2018).

Gore, A. (2007) *The Assault on Reason*. Bloomsbury, London, UK.

Grahame, K. (1908) *The Wind in the Willows*. Methuen, London, UK.

Grandin, T. (2003) Transferring results of behavioral research to industry to improve animal welfare on the farm, ranch and the slaughter plant. *Applied Animal Behaviour Science* 81, 215–228.

Grandin, T. (2013) Making slaughterhouses more humane for cattle, pigs, and sheep. *Annual Review of Animal Biosciences* 1, 491–512.

Grandin, T. (2015) *Improving Animal Welfare: A Practical Approach*, 2nd edn. CABI, Wallingford, UK.

Grandin, T. and Johnson, C. (2010) *Animals Make Us Human. Creating the Best Life for Animals*. Mariner Books, Boston, MA.

Gray, J. (2002) *Straw Dogs. Thoughts on Humans and Other Animals*. Granta Books, London, UK.

Green, T.C. and Mellor, D.J. (2011) Extending ideas about animal welfare assessment to include 'quality of life' and related concepts. *New Zealand Veterinary Journal* 59, 316–324.

Greene, J. (2013) *Moral Tribes. Emotion, Reason, and the Gap between Us and Them*. Atlantic, London, UK.

Gregory, N.G. (2004) *Physiology and Behaviour of Animal Suffering*. Blackwell, London, UK.

Gundersen, A.G. (1995) *The Environmental Promise of Democratic Deliberation*. University of Wisconsin, Madison, WI.

Haidt, J. (2012) *The Righteous Mind. Why Good People Are Divided by Politics and Religion*. Penguin, London, UK.

Hare, B., Brown, M., Williamson, C. and Tomasello, M. (2002) The domestication of social cognition in dogs. *Science* 298, 1634–1636.

Harford, T. (2016) *Messy. How to Be Creative and Resilient in a Tidy-Minded World*. Little, Brown, London, UK.

Harman, W.W. (1976) *An Incomplete Guide to the Future*. San Francisco Book Co., San Francisco, CA.

Harrison, R. (1964) *Animal Machines*. Vincent Stuart, London, UK (2013 edn. CABI, Wallingford, UK).

Harrison, R. (1971) On factory farming. In: Godlovitch, S., Godlovitch, R. and

Harris, J. (eds) *Animals, Men and Morals. An Enquiry into the Maltreatment of Non-Humans.* Victor Gollancz, London, UK, pp. 11–24.

Harrison, R. (1988) Special address. *Applied Animal Behaviour Science* 20, 21–27.

Harrison, R. (1991) The myth of the barn egg. *New Scientist*, 30 November, pp. 34–37.

Havenstein, G.B., Ferket, P.R. and Qureshi, M.A. (2003) Carcass composition and yield of 1957 versus 2001 broilers when fed representative 1957 and 2001 broiler diets. *Poultry Science* 82, 1509–1518.

Heap, R.B. (1995) Agriculture and bioethics – harmony or discord? *Journal of the Royal Agricultural Society of England* 156, 69–78.

Hemmer, H. (1990) *Domestication. The Decline of Environmental Appreciation.* Cambridge University Press, Cambridge, UK.

Hemsworth, P.H. (2004) Human-livestock interaction. In: Benson, G.E. and Rollin, B.E. (eds) *The Well-Being of Farm Animals. Challenges and Solutions.* Blackwell, Oxford, UK, pp. 21–38.

Hemsworth, P.H. and Gonyou, H.W. (1997) Human contact. In: Appleby, M.C. and Hughes, B.O. (eds) *Animal Welfare.* CAB International, Wallingford, UK, pp. 205–305.

Hendrickson, M.K. and James, H.S. (2005) The ethics of constrained choice: how the industrialization of agriculture impacts on farming and farmer behaviour. *Journal of Agricultural and Environmental Ethics* 18, 269–291.

Henrich, J., Heine, S.J. and Norenzayan, A. (2010a) Most people are not WEIRD. *Nature* 466, 1 July, p. 29.

Henrich, J., Heine, S.J. and Norenzayan, A. (2010b) The weirdest people in the world? *Behavioral and Brain Sciences* 33, 61–135.

Henshaw, D. (1989) *Animal Warfare. The Animal Liberation Front.* Fontana, London, UK.

Hinman, L. (1998) *Ethics. A Pluralistic Approach to Moral Theory.* Harcourt Brace, Fort Worth, TX.

Hobsbawm, E.J. and Rude, G. (1969) *Captain Swing.* Lawrence and Wishart, London, UK.

Hodges, J. (2000) Community of life. The ethical way forward. In: Hodges, J. and Han, I.K. (eds) *Livestock, Ethics and Quality of Life.* CAB International, Wallingford, UK, pp. 253–262.

Hodges, J. (2005) Cheap food and feeding the world sustainably. *Livestock Production Science* 92, 1–16.

Holmes, C. (1998) The Riddett Memorial Address. Dairy production in New Zealand 1948 to 1998; focus on systems for the future. Statistics for dairy farms. *Massey Dairyfarming Annual* 50, 11–30.

Holmes, C.W., Wilson, G.F., MacKenzie, D.D.S., Flux, D.S., Brookes, I.M. and Davey, A.W.F. (1987) *Milk Production from Pasture.* Butterworths, Wellington, NZ.

Hubbard, C. (2012) Do farm assurance schemes make a difference to animal welfare? *The Veterinary Record* 170, 150–151.

Humber, R.D. (1966) *Game Cock & Countryman.* Cassell, London, UK.

ITAVI (2003) Les syntheses economiques de l'ITAVI. Performances techniques et couts de production en volailles de chair, poluettes et poules pondeuses. Resultats 2002. ITAVI, France.

Jack, A. (2009) *Animal Welfare, Environmental, & Ethical Issues Affecting the Value of NZs Pastoral Products*. New Zealand Nuffield Scholarship Trust, Lincoln, NZ.

Jackson, T. (2009) *Prosperity without Growth. Economics for a Finite Planet*. Earthscan, Abingdon, UK.

Jacobsen, E. and Dulsrud, A. (2007) Will consumers save the world? The framing of political consumerism. *Journal of Agricultural and Environmental Ethics* 20, 469–482.

Jamieson, J., Reiss, M.J., Allen, D., Asher, L., Parker, M.O., Wathes, C.M. and Abeysinghe, S.M. (2015) Adolescents care but don't feel responsible for animal welfare. *Society and Animals* 23, 269–297.

Johnsen, P.F., Johannesson, T. and Sandøe, P. (2001) Assessment of farm animal welfare at herd level. *Acta Agriculturae Scandinavica, Section A, Animal Science* 51 (Suppl. 30), 26–33.

Johnson, A. (1991) *Factory Farming*. Blackwell, Oxford, UK.

Johnson, A. (1995) Barriers to the fair treatment of non-animal life. In: Cooper, D.E. and Palmer, J.A. (eds) *Just Environments. Intergenerational, International and Interspecies Issues*. Routledge, London, UK, pp. 165–179.

Jones, E.L. (1962) The changing basis of English agricultural prosperity, 1853–73. *The Agricultural History Review* 10, 102–119.

Jones, H. (2011) Taking responsibility for complexity: how implementation can achieve results in the face of complex problems. Briefing Paper 68, Overseas Development Institute, London. Available at: https://www.odi.org/publications/5275-complex-problems-complexity-implementation-policy (accessed 4 April 2018).

Jones, M. (2000) *The Tolpuddle Martyrs*. Trades Union Congress, London, UK.

Kasene, P. (1994) African ethical theory and the four principles. In: Gillon, R. (ed.) *Principles of Health Care Ethics*. John Wiley & Sons, Chichester, UK, pp. 183–192.

Kellert, S.R. and Wilson, E.O. (1993) *The Biophilia Hypothesis*. Island Press, Washington, D.C.

Kelly, R.W. and Whateley, J.A. (1975) Observations on the calving of red deer (Cervus elaphus) run in confined areas. *Applied Animal Behaviour Science* 1, 293–300.

Kestin, S.C., Knowles, T.G., Tinch, A.E. and Gregory, N.G. (1992) Prevalence of leg weakness in broiler chickens and its relationship with genotype. *The Veterinary Record* 131, 190–194.

Kilgour, R. (1972) Behaviour of sheep at lambing. *New Zealand Journal of Agriculture*, September, pp. 24–27.

Kilgour, R. (1975) The open-field test as an assessment of the temperament of dairy cows. *Animal Behaviour* 23, 615–624.

Kilgour, R. (1976) Cow behaviour. *NZ Farmer*, 10 June, Supplement, pp. 13–14.

Kilgour, R. (1978) The application of animal behavior and the humane care of farm animals. *Journal of Animal Science* 46, 1478–1486.

Kilgour, R. (1985) Animal welfare considerations – pastoral animals. *New Zealand Veterinary Journal* 33, 54–57.

Kilgour, R. and Dalton, C. (1984) *Livestock Behaviour. A Practical Guide.* Methuen, Auckland, NZ.

Kilgour, R. and Mullord, M. (1973) Transport of calves by road. *New Zealand Veterinary Journal* 21, 7–10.

Kjaer, J.B. and Mench, J.A. (2003) Behaviour problems associated with selection for increased production. In: Muir, W.M. and Aggrey, S.E. (eds) *Poultry Genetics, Breeding, and Biotechnology.* CABI, Wallingford, UK, pp. 67–82.

Klinenborg, V. and Modica, A. (2001) Cow parts. Mad cow disease could wreak havoc in the US because nearly everything we taste has cow in it. *Discover*, 1 August. Available at: http://discovermagazine.com/2001/aug/featcow (accessed 30 March 2018).

Knierim, U. and Jackson, W.T. (1997) Legislation. In: Appleby, M.C. and Hughes, B.O. (eds) *Animal Welfare.* CABI Publishing, Wallingford, UK, pp. 249–264.

Knight, T.W., Lynch, P.R., Hall, D.R.H. and Hockey, H.-U.P. (1988) Identification of factors contributing to the improved lamb survival in Marshall Romney sheep. *New Zealand Journal of Agricultural Research* 31, 259–271.

Knowles, T.G., Kestin, S.C., Haslam, S.M., Brown, S.N., Green, S.E., Butterworth, A., Pope, S.J., Pfeiffer, D. and Nicol, C.J. (2008) Leg disorders in broiler chickens: prevalence, risk factors and prevention. *PLoS ONE* 3 (2), 1–5.

Korthals, M. (2008) Ethical rooms for maneuver and their prospects vis-à-vis the current ethical food policies in Europe. *Journal of Agricultural and Environmental Ethics* 21, 249–273.

Lagerkvist, C.J. and Hess, S. (2011) A meta-analysis of consumer willingness to pay for farm animal welfare. *European Review of Agricultural Economics* 38, 55–78.

Lang, T. and Barling, D. (2013) UK food policy. Can we get it on the right track? *Food Ethics Council Bulletin* 8 (3), 4–6.

Lang, T. and Heasman, M. (2004) *Food Wars. The Global Battle for Mouths, Minds and Markets.* Earthscan, London, UK.

Lara, L.J. and Rostagno, M.H. (2018) Animal welfare and food safety in modern animal production. In: Mench, J.A. (ed.) *Advances in Agricultural Animal Welfare. Science and Practice.* Woodhead Publishing, Duxford, UK, pp. 91–108.

Lassen, J., Gjerris, M. and Sandøe, P. (2006a) After Dolly—ethical limits to the use of biotechnology on farm animals. *Theriogenology* 65, 992–1004.

Lassen, J., Sandøe, P. and Forkman, B. (2006b) Happy pigs are dirty!: conflicting perspectives on animal welfare. *Livestock Science* 103, 221–230.

Laundré, J. (2016) Predation. In: Villalba, J.J. (ed.) *Animal Welfare in Extensive Production Systems.* 5M, Sheffield, UK, pp. 103–131.

Lawler, A. (2016) *How the Chicken Crossed the World. The Story of the Bird That Powers Civilizations.* Duckworth Overlook, London, UK.

Lawrence, A.B., Conington, J. and Simm, G. (2004) Breeding and animal welfare: practical and theoretical advantages of multi-trait selection. *Animal Welfare* 13 (Supplement 1), 191–196.

Lawrence, E.A. (1993) The sacred bee, the filthy pig, and the bat out of hell:

animal symbolism as cognitive biophilia. In: Kellert, S.R. and Wilson, E.O. (eds) *The Biophilia Hypothesis.* Island Press, Washington, D.C., pp. 301–341.

Le Guin, U.K. (1976) *The Wind's Twelve Quarters. Short Stories.* Victor Gollancz, London, UK.

Lebas, F., Coudert, P., de Rochambeau, H. and Thébault, R.G. (1997) *The Rabbit. Husbandry, Health and Production.* FAO, Rome, Italy.

Leibler, J.H., Otte, J., Roland-Holst, D., Pfeiffer, D.U., Soares Magalhaes, R., Rushton, J., Graham, J.P. and Silbergeld, E.K. (2009) Industrial food animal production and global health risks: exploring the ecosystems and economics of avian influenza. *Ecohealth* 6, 58–70.

Leopold, A. (1949) *A Sand County Almanac. And Sketches Here and There.* Oxford University Press, New York, NY.

Lescourret, F., Coulon, J.B. and Faye, B. (1995) Predictive model of mastitis occurrence in the dairy cow. *Journal of Dairy Science* 78, 2167–2177.

Levis, D.G. (2016) The modern sow: top production issues. In: *Proceedings of the London Swine Conference: A Platform for Success, April 5–6, 2016.* London Swine Conference, London, ON, Canada, pp. 19–34.

Lewis, J. (2011) Young carnivores should be taught the truth. *The Telegraph*, 1 December. Available at: http://www.telegraph.co.uk/foodanddrink/8928988/Young-carnivores-should-be-taught-the-truth.html (accessed 31 March 2018).

Lobao, L. and Stafferahn, C.W. (2005) The community effects of industrialized farming: social science research and challenges to corporate farming laws. *Agriculture and Human Values* 25, 219–240.

Lockwood, J.A. (2002) *Grasshopper Dreaming. Reflections on Killing and Loving.* Skinner House, Boston, MA.

Lowman, B. (1999) The producer's view – ruminants. In: Russell, A.J.F., Morgan, C.A., Savory, C.J., Appleby, M.C. and Lawrence, T.L.J. (eds) *Farm Animal Welfare – Who Writes The Rules? British Society of Animal Production Occasional Publication* 23, 15–18.

Lubchenco, J. (1998) Entering the century of the environment: a new social contract for science. *Science* 279, 491–497.

Lucy, M.C. (2001) Reproductive physiology and management of high-yielding dairy cattle. *Proceedings of the New Zealand Society of Animal Production* 61, 120–127.

Lund, B. (2005) *Goose Fair on Old Picture Postcards.* Reflections of a Bygone Age, Nottingham, UK.

Lundmark, F., Berg, C. and Röcklinsberg, H. (2018) Private animal welfare standards—opportunities and risks. *Animals* 8, 4.

Lusk, J.L. and Norwood, F.B. (2008) A survey to determine public opinion about the ethics and governance of farm animal welfare. *Journal of the American Veterinary Medical Association* 233, 1121–1126.

Lusk, J.L. and Norwood, F.B. (2011a) A calibrated auction-conjoint valuation method: valuing pork and eggs produced under differing animal welfare conditions. *Journal of Environmental Economics and Management* 62, 80–94.

Lusk, J.L. and Norwood, F.B. (2011b) Animal welfare economics. *Applied Economic Perspectives and Policy* 33, 463–483.

Lusk, J.L. and Norwood, F.B. (2012) Speciesism, altruism and the economics of animal welfare. *European Review of Agricultural Economics* 39, 189–212.

Lymbery, P. and Oakeshott, I. (2014) *Farmageddon. The True Cost of Cheap Meat.* Bloomsbury, London, UK.

MacLean, H. (2016) 'Inability to control his frustration'. *Otago Daily Times*, 3 November. Available at: https://www.odt.co.nz/regions/north-otago/inability-control-his-frustration (accessed 31 March 2018).

MacLeod, C.J. and Moller, H. (2006) Intensification and diversification of New Zealand agriculture since 1960: an evaluation of current indicators of land use change. *Agriculture, Ecosystems & Environment* 115, 201–218.

McCorry, S. (2013) Agrarian nostalgia and industrial agriculture: George Orwell's Animal Farm as political pastoral. In: Calvert, C. and Gröling, J. (eds) *Proceedings of the Conference Critical Perspectives on Animals in Society, University of Exeter, UK*, pp. 31–36.

McCulloch, S.P. and Reiss, M.J. (2017) Bovine tuberculosis and badger control in Britain: science, policy and politics. *Journal of Agricultural and Environmental Ethics* 30, 469–484.

McDougall, S. (2006) Reproductive performance and management of dairy cattle. *Journal of Reproduction and Development* 52, 185–194.

McGregor, B.A. and Butler, K.L. (2008) Relationship of body score, live weight, stocking rate and grazing system to the mortality of Angora goats from hypothermia and their use in the assessment of welfare risks. *Australian Veterinary Journal* 86, 12–17.

McInerney, J. (2004) Animal welfare, economics and policy. Available at: http://webarchive.nationalarchives.gov.uk/20110318142209/http://www.defra.gov.uk/evidence/economics/foodfarm/reports/documents/animalwelfare.pdf (accessed 31 March 2018).

McInerney, J. (2013) Principles, preference and profit: animal ethics in a market economy. In: Wathes, C.M., Corr, S.A., May, S.A., McCulloch, S.P. and Whiting, M.C. (eds) *Veterinary & Animal Ethics: Proceedings of the First International Conference on Veterinary and Animal Ethics, September 2011.* Wiley-Blackwell, Oxford, UK, pp. 271–285.

McKenna, E. and Light, A. (2004) *Animal Pragmatism. Rethinking Human-Nonhuman Relationships.* Indiana University, Indianapolis, IN.

McKibben, B. (2003) *The End of Nature. Humanity, Climate Change and the Natural World.* Bloomsbury, London, UK.

McLean, W.H. (1966) *Rabbits Galore! On the Other Side of the Fence.* A.H. & A.W. Reed, Wellington, NZ.

McMillan, F.D. (2000) Quality of life in animals. *Journal of the American Veterinary Medical Association* 216, 1904–1910.

Mason, G.J. and Latham, N.R. (2004) Can't stop, won't stop: is stereotype a reliable animal welfare indicator? *Animal Welfare* 13, S57–69.

Matheny, G. and Leahy, C. (2007) Farm-animal welfare, legislation, and trade. *Law and Contemporary Problems* 70, 325–358.

Mazoyer, M. and Roudart, L. (2006) *A History of World Agriculture. From the Neolithic Age to the Current Crisis.* Earthscan, Abingdon, UK.

Meaden, D., Darwent, N. and Lundgren, P. (undated) *Weighing up the Economics of Dairy Farms.* A briefing by the World Society for the Protection of Animals. World Society for the Protection of Animals, London, UK.

Mellor, D.J. and Beausoleil, N.J. (2015) Extending the 'five domains' model for animal welfare assessment to incorporate positive welfare states. *Animal Welfare* 24, 241–253.

Mellor, D.J., Diesch, T.J. and Johnson, C.B. (2010) Should mammalian fetuses be excluded from regulations protecting animals during experiments? *ALTEX* 27, Special Issue, 199–202.

Mellor, D.J., Patterson-Kane, E. and Stafford, K.J. (2009) *The Sciences of Animal Welfare.* Wiley-Blackwell, Oxford, UK.

Mellor, D.J. and Reid, C.S. (1994) Concepts of animal well-being and predicting the impact of procedures on experimental animals. In: Baker, R., Jenkin, G. and Mellor, D.J. (eds) *Improving the Well-Being of Animals in the Research Environment.* ANZCCART, Glen Osmond, Australia, pp. 3–18.

Mepham, B. (2000) A framework for the ethical analysis of novel foods: the ethical matrix. *Journal of Agricultural and Environmental Ethics* 12, 165–176.

Mepham, B. (2001) *Farming Animals for Food: Towards a Moral Menu.* Food Ethics Council, London, UK.

Mepham, T.B. and Forbes, J.M. (1995) Ethical aspects of the use of immunomodulation in farm animals. *Livestock Science* 42, 265–272.

Meyfroidt, P., Lambin, E.F., Erb, K.-H. and Hertel, T.W. (2013) Globalization of land use: distant drivers of land change and geographic displacement of land use. *Current Opinion in Environmental Sustainability* 5, 1–7.

Midgley, M. (1983) *Animals and Why They Matter.* University of Georgia Press, Athens, GA.

Midgley, M. (1991) *Wisdom, Information and Wonder.* Routledge, London, UK.

Midgley, M. (1993) The origin of ethics. In: Singer, P. (ed.) *A Companion to Ethics.* Blackwell, Oxford, UK, pp. 3–13.

Millman, S.T., Duncan, I.J.H. and Widowski, T.M. (2000) Male broiler breeder fowl display high levels of aggression toward females. *Poultry Science* 79, 1233–1241.

Milner, R. (2015) Wagyu: processing pampered cows at Tokyo's last major slaughterhouse. *The Japan Times*, 7 August. Available at: https://www.japantimes.co.jp/life/2015/08/07/food/wagyu-processing-pampered-cows-tokyos-last-major-slaughterhouse/ (accessed 4 April 2018).

Misculin, N. (2008) Argentine cattle move from Pampas to feedlots. *Reuters*, 18 December. Available at: https://www.reuters.com/article/us-argentina-feedlots/argentine-cattle-move-from-pampas-to-feedlots-idUSTRE4BH02B20081218 (accessed 7 April 2018).

Mithen, S. (1996) *The Prehistory of the Mind. A Search for the Origins of Art, Religion and Science.* Phoenix, London, UK.

Monbiot, G. (2012) If children lose contact with nature they won't fight for it. *The Guardian*, 19 November. Available at: https://www.theguardian.com/commentisfree/2012/nov/19/children-lose-contact-with-nature (accessed 4 April 2018).

Monbiot, G. (2016) *How Did We Get in to this Mess? Politics, Equality and Nature*. Verso, London, UK.

Monbiot, G. (2017) *Out of the Wreckage. A New Politics for an Age of Crisis*. Verso, London, UK.

Monnerot, M., Vigne, J.D., Biju-Duval, C., Casane, D., Callou, C., Hardy, C., Mougel, F., Soriguer, R., Dennebouy, N. and Mounolou, J.C. (1994) Rabbit and man: genetic and historic approach. *Genetics Selection Evolution* 26, 167s–182s.

More, S.J., Hanlon, A., Marchewka, J. and Boyle, L. (2017) Private animal health and welfare standards in quality assurance programmes: a review and proposed framework for critical evaluation. *The Veterinary Record* 180, 612.

More, T. (1516) *Utopia*. Risk, Ewing and Smith, Dublin, Ireland.

Moreira, J.R., Alvarez, M.R., Tarifa, T., Pacheco, V., Taber, A., Tirira, D.G., Herrera, E.A., Ferraz, K.M.P.M.B., Aldana-Domínguez, J. and Macdonald, D.W. (2013) Taxonomy, natural history and distribution of the capybara. In: Moreira, J.R., Ferraz, K.M.P.M.B., Herrera, E.A. and Macdonald, D. (eds) *Capybara: Biology, Use and Conservation of an Exceptional Neotropical Species*. Springer, New York, NY, pp. 3–37.

Morrison, J.M. (1998) The poultry industry: a view of the swine industry's future? In: Thu, K.M. and Durrenberger, E.P. (eds) *Pigs, Profits, and Rural Communities*. State University of New York Press, Albany, NY, pp. 145–154.

Mortimer, J. (2018) Kiwi sausage company grilled over 'foul, disgusting' image. *New Zealand Herald*, 15 January. Available at: http://www.nzherald.co.nz/lifestyle/news/article.cfm?c_id=6&objectid=11975245 (accessed 31 March 2018).

Mulligan, F.J. and Doherty, M.L. (2008) Production diseases of the transition cow. *The Veterinary Journal* 176, 3–9.

Nasr, M.A.F., Murrell, J., Wilkins, L.J. and Nicol, C.J. (2012) The effect of keel fractures on egg-production parameters, mobility and behaviour in individual laying hens. *Animal Welfare* 21, 127–135.

Nelson, R. (1993) Searching for the lost arrow: physical and spiritual ecology in the hunter's world. In: Kellert, S.R. and Wilson, E.O. (eds) *The Biophilia Hypothesis*. Island Press, Washington, D.C., pp. 201–223.

Newberry, R.C. and Sandilands, V. (2016) Pioneers of applied ethology. In: Brown, J.A., Seddon, Y.M. and Appleby, M.C. (eds) *Animals and Us: 50 Years and More of Applied Ethology*. Wageningen Academic, Wageningen, the Netherlands, pp. 51–75.

Niamir-Fuller, M. (2016) Towards sustainability in the extensive and intensive livestock sectors. *Revue Scientifique et Technique de l'Office International des Épizooties* 35, 371–387.

Nickles, B. (1998) An Iowa farmer's personal and political experience with factory hog facilities. In: Thu, K.M. and Durrenberger, E.P. (eds) *Pigs, Profits, and Rural Communities*. State University of New York Press, Albany, NY, pp. 123–137.

Nielsen, B.L. (1999) Perceived welfare issues in dairy cattle, with special emphasis on metabolic stress. In: Oldhorn, J.D., Simm, G., Groen, A.F., Nielsen, B.L., Pryce, J.E. and Lawrence, T.L.J. *Metabolic Stress in Dairy Cows. British Society of Animal Science Occasional Publication* 24, 1–7.

Niles, M.T. (2013) Achieving social sustainability in animal agriculture: challenges and opportunities to reconcile multiple sustainability goals. In: Kebreab, E. (ed.) *Sustainable Animal Agriculture*. CAB International, Wallingford, UK, pp. 193–211.

Nir, O. (2003) What are production diseases and how do we manage them? *Acta Veterinaria Scandinavica Supplementum* 98, 21–32.

Noe, E. and Alrøe, H.F. (2003) Farm enterprises as self-organizing systems: a new transdisciplinary framework for studying farm enterprises? *International Journal of Sociology of Agriculture and Food* 11, 3–14.

Nogueira, S.S.C. and Nogueira-Filho, S.L.G. (2012) Capybara (Hydrochoerus hydrochaeris) behaviour and welfare: implications for successful farming practices. *Animal Welfare* 21, 527–533.

Nordenfelt, L. (2006) *Animal and Human Health and Welfare. A Comparative Philosophical Analysis*. CABI, Wallingford, UK.

Norman, J. (2013) *Edmund Burke. The Visionary Who Invented Modern Politics*. William Collins, London, UK.

Norwood, F.B. and Lusk, J.L. (2011) *Compassion by the Pound. The Economics of Farm Animal Welfare*. Oxford University Press, Oxford, UK.

Nussbaum, M.C. (2004) Beyond "compassion and humility": justice for nonhuman animals, In: Sunstein, C.R. and Nussbaum, M.C. (eds) *Animal Rights: Current Debates and New Directions*. Oxford University Press, Oxford, UK, pp. 299–320.

Olsson, I.A.S. and Keeling, L.J. (2002) The push-door for measuring motivation in hens: laying hens are motivated to perch at night. *Animal Welfare* 11, 11–19.

Orquera, L.A. (1984) Specialization and the Middle/Upper Paleolithic transition. *Current Anthropology* 25, 73–98.

Orwell, G. (1945) *Animal Farm. A Fairy Story*. Penguin, London, UK.

Ostrom, E. (2009) A general framework for analysing sustainability of social-ecological systems. *Science* 325, 419–422.

Ott, R.S. (1996) Animal selection and breeding techniques that create diseased populations and compromise welfare. *Journal of the American Veterinary Medical Association* 208, 1969–1974.

Overton, M. (1996) *Agricultural Revolution in England. The Transformation of the Agrarian Economy 1500–1850*. Cambridge University Press, Cambridge, UK.

Ozeki, R.L. (1998) *My Year of Meat*. Pan Books, London, UK.

Pacelle, W. (2016) *The Humane Economy. How Innovators and Enlightened Consumers Are Transforming the Lives of Animals*. William Morrow, New York, NY.

Palmer, Sir G. (2015) Climate change and New Zealand. Is it doom or can we cope? *Policy Quarterly* 11, 15–24.

Parliamentary Commissioner for the Environment (1998) *The Rabbit Calicivirus Disease (RCD) Saga: A Biosecurity/Biocontrol Fiasco*. Office of the Parliamentary Commissioner for the Environment, Wellington, NZ.

Parliamentary Commissioner for the Environment (2004) *Growing for Good: Intensive Farming, Sustainability and New Zealand's Environment*. Office of the Parliamentary Commissioner for the Environment, Wellington, NZ.

Parminter, T.G. and Perkins, A.M.L. (1997) Applying an understanding of farmers' values and goals to their farming styles. *Proceedings of the New Zealand Grassland Association* 59, 107–111.

Passarello, E. (2017) *Animals Strike Curious Poses. Essays*. Jonathan Cape, London, UK.

Patterson, C. (2002) *Eternal Treblinka. Our Treatment of Animals and the Holocaust*. Lantern Books, New York, NY.

Pawson, H.C. (1957) *Robert Bakewell. Pioneer Livestock Breeder*. Crosby Lockwood & Son, London, UK.

Payne, J.M. (1972) Production disease. *Journal of the Royal Agricultural Society of England* 133, 69–86.

Perdrizet, G.A. (1997) Hans Selye and beyond: responses to stress. *Cell Stress & Chaperones* 2, 214–219.

Perry, P.J. (1981) High farming in Victorian Britain: prospect and retrospect. *Agricultural History* 55, 156–166.

Peter, M.L. (2010) How to argue for agriculture—and how not to. *Angus Beef Bulletin Extra*, 10 March. Available at: http://www.angusbeefbulletin.com/extra/2010/03mar10/0310fp_argue4.html (accessed 7 April 2018).

Peterson, H.C. (2013) Sustainability: a wicked problem. In: Kebreab, E. (ed.) *Sustainable Animal Agriculture*. CAB International, Wallingford, UK, pp. 1–9.

Petherick, J.C. (2005) Animal welfare issues associated with extensive livestock production: the northern Australian beef cattle industry. *Applied Animal Behaviour Science* 92, 211–234.

Petrini, A. and Wilson, D. (2005) Philosophy, policy and procedures of the World Organisation for Animal Health for the development of standards in animal welfare. Animal welfare: global issues, trends and challenges. *Revue Scientifique et Technique de l'Office International des Épizooties* 24, 665–671.

Petticrew, M.P. and Lee, K. (2011) The "Father of Stress" meets "Big Tobacco": Hans Selye and the tobacco industry. *American Journal of Public Health* 101, 411–418.

Pew Commission (2008) Putting meat on the table: industrial farm animal production in America. Available at: http://www.pewtrusts.org/-/media/legacy/uploadedfiles/phg/content_level_pages/reports/pcifapfinalpdf.pdf (accessed 14 June 2018).

Pielke, R.A. Jr (2007) *The Honest Broker. Making Sense of Science in Policy and Politics*. Cambridge University Press, Cambridge, UK.

Pike, A.W.G., Gilmour, M., Petitt, P., Jacobi, R., Ripoll, S., Bahn, P. and Muñoz, F. (2005) Verification of the age of the Palaeolithic cave art at Creswell Crags, UK. *Journal of Archaeological Science* 32, 1649–1655.

Piper, H.B. (1962) *Little Fuzzy*. Avon, NY (2015 edn, Jefferson Publication).
Plain, R.L. (2010) Historical perspective of the integration of animal agriculture. In: Reynnells, R. and Chimenti, L.M. (eds) *Sustaining Animal Agriculture: Balancing Bioethical, Economic, and Social Issues*. CAST Food-Animal Agriculture Symposium, CAST and USDA, Washington D.C., pp. 29–33 and 316–408. Available at: https://permanent.access.gpo.gov/gpo3872/Proceedings-Final-for-web.pdf (accessed 7 April 2018).
Podbersccek, A.L. (2009) Good to pet and eat: the keeping and consuming of dogs and cats in South Korea. *Journal of Social Issues* 65, 615–632.
Pollan, M. (2002) An animal's place. *New York Times Magazine*, 10 November, pp. 58–64.
Polyani, K., Arensberg, C.M. and Pearson, H.W. (1957) *Trade and Market in the Early Empires. Economies in History and Theory*. Free Press, New York, NY.
Pösö, J. and Mäntysaari, E.A. (1996) Relationships between clinical mastitis, somatic cell score, and production for the first three lactations of Finnish Ayrshire. *Journal of Dairy Science* 79, 1284–1291.
Potter, V.R. (1970) Bioethics, the science of survival. *Perspectives in Biology and Medicine*, Autumn, pp. 127–153.
Potter, V.R. (1971) *Bioethics. Bridge to the Future*. Prentice-Hall, Englewood Cliffs, NJ.
Potter, V.R. (1988) *Global Bioethics. Building on the Leopold Legacy*. Michigan State, East Lansing, MI.
Potter, V.R. (1990) Getting to the year 3000: can global bioethics overcome evolution's fatal flaw? *Perspectives in Biology and Medicine* 34, 89–98.
Pradère, J.-P. (2017) *Poor Livestock Producers, the Environment and the Paradoxes of Development Policies*. World Organisation for Animal Health (OIE), Paris, France.
Pratchett, T. (2005) *Where's My Cow?* Doubleday, London, UK.
Preece, R. (1999) *Animals and Nature: Cultural Myths, Cultural Realities*. UBC Press, Vancouver, Canada.
Preece, R. (2002) *Awe for the Tiger, Love for the Lamb. A Chronicle of Sensibility to Animals*. Routledge, New York, NY.
Presland, P. (2011) When co-dependency breaks down. *VetScript*, March, pp. 10–12.
Pretty, J. (2002) *Agri-Culture. Reconnecting People, Land and Nature*. Earthscan, London, UK.
Pretty, J. (2007) *The Earth Only Endures. On Reconnecting with Nature and Our Place in it*. Earthscan, London, UK.
Price, E.O. (1984) Behavioral aspects of animal domestication. *Quarterly Review of Biology* 59, 1–32.
Price, G. (1874) *The Skeleton at the Plough*. Potter, London, UK.
Prickett, R.W., Norwood, F., Bailey, F. and Lusk, J.L. (2010) Consumer preferences for farm animal welfare: results from a telephone survey of US households. *Animal Welfare* 19, 335–347.
Protopopova, A. and Gunter, L.M. (2017) Adoption and relinquishment interventions at the animal shelter: a review. *Animal Welfare* 26, 35–48.

Pryce, J.E., Parker Gaddis, K.L., Koeck, A., Bastin, C., Abdelsayed, M., Gengler, N., Miglior, F., Heringstad, B., Egger-Danner, C., Stock, K.F., Bradley, A.J. and Cole, J.B. (2016) Opportunities for genetic improvement of metabolic diseases. *Journal of Dairy Science* 99, 6855–6873.

Pryce, J.E., Veerkamp, R.F., Thompson, R., Hill, W.G. and Simm, G. (1997) Genetic aspects of common health disorders and measures of fertility in Holstein Friesian dairy cattle. *Animal Science* 65, 353–360.

Putu, I.G., Poindron, P. and Lindsay, D.R. (1988) Early disturbance of Merino ewes from the birth site increases lamb separation and mortality. *Proceedings of the Australian Society of Animal Production* 17, 298–301.

Queck, P. (2013) Argentina provides a lesson in how to ruin a beef industry. *Beef*, 26 September. Available at: http://www.beefmagazine.com/beef-exports/argentina-provides-lesson-how-ruin-beef-industry (accessed 7 April 2018).

Rachels, J. (1990) *Created from Animals. The Moral Implications of Darwinism.* Oxford University Press, Oxford, UK.

Rahman, S.A. (2017) Religion and animal welfare—an Islamic perspective. *Animals* 7, 11.

Rauw, W.M. (2015) Philosophy and ethics of animal use and consumption: from Pythagoras to Bentham. *CAB Reviews* 10 (16), 1–25.

Rauw, W.M., Kannis, E., Noordhuizen-Stassen, E.N. and Grammers, F.J. (1998) Undesirable side effects of selection for high production efficiency in farm animals: a review. *Livestock Production Science* 56, 15–33.

Rebanks, J. (2015) *The Shepherd's Life: A Tale of the Lake District.* Penguin, London, UK.

Regan, T. (1983) *The Case for Animal Rights.* Routledge, London, UK.

Reimer, T., Dempster, T., Wargelius, A., Gunnar Fjelldal, P., Hansen, T., Glover, K.A., Solberg, M.F. and Swearer, S.E. (2017) Rapid growth causes abnormal vaterite formation in farmed fish otoliths. *Journal of Experimental Biology* 220, 2965–2969.

Reimer, T., Dempster, T., Warren-Myers, F., Jensen, A.J. and Swearer, S.E. (2016) High prevalence of vaterite in sagittal otoliths causes hearing impairment in farmed fish. *Nature Scientific Reports* 6, 25249.

Reynnells, R. (2007) Proactive approaches to controversial welfare and ethical concerns in poultry science. USDA, Washington, D.C.

Ricard, R. (2015) The natural repugnance to kill. Chapter 29 in *Altruism: The Power of Compassion to Change Yourself and the World.* Atlantic Books, London, UK.

Ricard, M. (2016) *A Plea for the Animals. The Moral, Philosophical, and Evolutionary Imperative to Treat all Beings with Compassion.* Shambala, Boulder, CO.

Richards, E., Signal, T. and Taylor, N. (2013) A different cut? Comparing attitudes toward animals and propensity for aggression within two primary industry cohorts—farmers and meatworkers. *Society & Animals* 21, 395–413.

Ridley, M. (2015) *The Evolution of Everything. How New Ideas Emerge.* Fourth Estate, London, UK.

Ritson, C. (2008) Food prices: what are they? *Food Ethics* 3 (2), 14–15.

Rittel, H.W.J. and Webber, M.M. (1973) Dilemmas in a general theory of planning. *Policy Sciences* 4, 155–169.

Ritzer, G. (2004) *The McDonaldization of Society*. Pine Forge Press, Thousand Oaks, CA.

Rizzi, M., Heath, M., Marsh, S. and Osterman, C. (2016) Argentina lifts beef export quotas, agriculture secretary says. *Reuters*, 11 January. Available at: https://www.reuters.com/article/argentina-beef/argentina-lifts-beef-export-quotas-agriculture-secretary-says-idUSL8N14V4NC20160111 (accessed 7 April 2018).

Robert, S., Weary, D.M. and Gonyou, H. (1999) Segregated early weaning and welfare of piglets. *Journal of Applied Animal Welfare Science* 2, 31–40.

Roberts, P. (2014) *The Impulse Society. What's Wrong with Getting What We Want?* Bloomsbury, London, UK.

Robinson, K.S. (2012) *2312*. Orbit, London, UK.

Roche, J.R., Berry, D.P., Bryant, A.M., Burke, C.R., Butler, S.T., Dillon, P.G., Donaghy, D.J., Horan, B.A., Macdonald, K.A. and Macmillan, K.L. (2017) A century of change in temperate grazing dairy systems. *Journal of Dairy Science* 100, 10189–10233.

Röcklinsberg, H., Gamborg, C. and Gjerris, M. (2014) A case for integrity: gains from including more than animal welfare in animal ethics committee deliberations. *Laboratory Animals* 48, 61–71.

Rodd, R. (1990) *Biology, Ethics, and Animals*. Clarendon Press, Oxford, UK.

Rogoff, M.H. and Rawlins, S.L. (1987) Food security: a technological alternative. *BioScience* 37, 800–807.

Roguski, M. (2012) *Pets as pawns: the coexistence of animal cruelty and family violence*. Report prepared for the Royal New Zealand Society for the Prevention of Cruelty to Animals and The National Collective of Independent Women's Refuges. Available at: www.communityresearch.org.nz/wp-content/uploads/formidable/Pets-as-Pawns-Research-Report-Final.pdf (accessed 7 April 2018).

Röling, N. (2000) Gateway to the global garden: beta/ gamma science for dealing with ecological rationality. Eighth Annual Hopper Lecture, University of Guelph, Canada. In: Pretty, J. (ed.) *The Earthscan Reader in Sustainable Agriculture*. Routledge, London, UK, pp. 179–205.

Rollin, B.E. (1990) Animal welfare, animal rights and agriculture. *Journal of Animal Science* 68, 3456–3461.

Rollin, B.E. (1992) *Animal Rights & Human Morality*. Prometheus, Buffalo, NY.

Rollin, B.E. (1993) Animal production and the new social ethic for animals. In: *Proceedings of the Food Animal Well-Being Conference and Workshop, Indianapolis, April 13-15, 1993*. USDA and Purdue University, West Lafayette, IN, pp. 3–13.

Rollin, B.E. (1995) *Farm Animal Welfare. Social, Bioethical, and Research Issues*. Iowa State University Press, Ames, IA.

Rollin, B.E. (1996) Bad ethics, good ethics and the genetic engineering of animals in agriculture. *Journal of Animal Science* 74, 535–541.

Rollin, B.E. (2006) Ethics, biotechnology and animals. In: Waldau, P. and Patton,

K. (eds) *A Communion of Subjects: Animals in Religion, Science, and Ethics*. Columbia University Press, New York, NY, pp. 519–532.

Rollin, B.E. (2008) The ethics of agriculture: the end of true husbandry. In: Dawkins, M.S. and Bonney, R. (eds) *The Future of Animal Farming. Renewing the Ancient Contract*. Blackwell, Oxford, UK, pp. 7–19.

Romanes, G.J. (1882) *Animal Intelligence*. Keegan, Paul, Trench & Co., London, UK.

Rose, D.B. (2000) *Dingo Makes Us Human. Life and Land in an Australian Aboriginal Culture*. Cambridge University Press, Cambridge, UK.

Roser, M. and Ritchie, H. (2018) Food prices. *Our World in Data*. Available at: https://ourworldindata.org/food-prices (accessed 7 April 2018).

Rushen, J. (2003) Changing concepts of farm animal welfare: bridging the gap between applied and basic research. *Applied Animal Behaviour Science* 81, 199–214.

Rutherford, K.M.D., Baxter, E.M., D'Eath, R.B., Turner, S.P., Arnott, G., Roehe, R., Ask, B., Sande, P., Mousten, V.A., Thorup, F., Edwards, S.A., Berg, P. and Lawrence, A.B. (2013) The welfare implications of large litter size in the domestic pig I: biological factors. *Animal Welfare* 22, 199–218.

Ryder, R.D. (1970) Speciesism. In: Speciesism again: the original leaflet. *Critical Society* 2, Spring 2010, 2 pp. Available at: https://web.archive.org/web/20121114004403/http://www.criticalsocietyjournal.org.uk/Archives_files/1.%20Speciesism%20Again.pdf (accessed 24 March 2018).

Ryder, R.D. (2000) *Animal Revolution. Changing Attitudes towards Speciesism*. Berg, Oxford, UK.

Rymer, R. (2012) Vanishing voices. The compassion of Khoj Özeeri. *National Geographic*, July. Available at: https://www.nationalgeographic.com/magazine/2012/07/vanishing-languages/ (accessed 14 June 2018).

Sachs, J. (2012) *The Price of Civilisation*. Vintage, London, UK.

Salt, H. (1892) *Animal Rights: Considered in Relation to Social Progress*. Centaur Press, London, UK (1980 edn).

Sandilands, V. (2011) The laying hen and bone fractures. *The Veterinary Record*, 15 October, pp. 411–412.

Sandøe, P. and Christiansen, S.B. (2008) *Ethics of Animal Use*. Wiley-Blackwell, Oxford, UK.

Sandøe, P., Christiansen, S.B. and Appleby, M.C. (2003) Farm animal welfare: the interaction of ethical questions and animal welfare science. *Animal Welfare* 12, 469–478.

Sandøe, P. and Jensen, K.K. (2013) The idea of animal welfare – developments and tensions. In: Wathes, C.M., Corr, S.A., May, S.A., McCulloch, S.P. and Whiting, M.C. (eds) *Veterinary & Animal Ethics. Proceedings of the First International Conference on Veterinary and Animal Ethics, September 2011*. Wiley-Blackwell, Oxford, UK, pp. 19–31.

Sandøe, P., Nielsen, B.L., Christiansen, L.G. and Sørensen, P. (1999) Staying good while playing God – the ethics of breeding farm animals. *Animal Welfare* 8, 313–328.

Sandøe, P. and Simonsen, H.B. (1992) Assessing animal welfare: where does science end and philosophy begin? *Animal Welfare* 1, 257–267.

Sato, P., Hötzel, M.J. and von Keyserlingk, M.A.G. (2017) American citizens' views of an ideal pig farm. *Animals* 7, 64.

Saul, J.R. (1992) *Voltaire's Bastards. The Dictatorship of Reason in the West*. Penguin, Toronto, Canada.

Saul, J.R. (1995) *The Unconscious Civilization*. Penguin, Ringwood, Australia.

Saul, J.R. (2001) *On Equilibrium*. Penguin, Ringwood, Australia.

Saul, J.R. (2005) *The Collapse of Globalism and the Reinvention of the World*. Penguin, Camberwell, Australia.

Scarlett, J.M., Salman, M.D., New, J.G. Jr and Kass, P.H. (1999) Reasons for relinquishment of companion animals in U.S. animal shelters: selected health and personal issues. *Journal of Applied Animal Welfare Science* 2, 41–57.

Schwartz, S. (2003) Separation anxiety syndrome in dogs and cats. *Journal of the American Veterinary Medical Association* 222, 1526–1532.

Schweitzer, A. (1936) The ethics of reverence for life. *Christendom* 1, 225–239.

Schweitzer, A. (1987) *The Philosophy of Civilisation*. Prometheus, New York, NY.

Scobie, D.R., Bray, A.R. and O'Connell, D. (1999) A breeding goal to improve the welfare of sheep. *Animal Welfare* 8, 391–406.

Scobie, D.R., O'Connell, D., Morris, C.A. and Hickey, S.M. (2007) A preliminary genetic analysis of breech and tail traits with the aim of improving the welfare of sheep. *Australian Journal of Agricultural Research* 58, 161–167.

Scully, M. (2002) *Dominion. The Power of Man, the Suffering of Animals, and the Call to Mercy*. St Martin's Griffin, New York, NY.

Selye, H. (1936) A syndrome produced by diverse nocuous agents. *Nature*, 4 July, p. 32.

Senge, P.M. (1990) *The Fifth Discipline. The Art and Practice of the Learning Organization*. Century Business, London, UK.

Serpell, J. (1986) *In the Company of Animals. A Study of Human–Animal Relationships*. Cambridge University Press, Cambridge, UK.

Setchell, B.P. (1992) Domestication and reproduction. *Animal Reproduction Science* 28, 195–202.

Sewell, A. (1877) *Black Beauty*. Jarrold & Sons, Norwich, UK.

Sheail, J. (1971) *Rabbits and Their History*. David & Charles, Newton Abbot, UK.

Sherman, B.L.I. and Mills, D.S. (2008) Canine anxieties and phobias: an update on separation anxiety and noise aversions. *Veterinary Clinics: Small Animal Practice* 38, 1081–1106.

Shields, S., Shapiro, S. and Rowan, A. (2017) A decade of progress toward ending the intensive confinement of farm animals in the United States. *Animals* 7, 40.

Shim, M.Y., Karnuah, A.B., Mitchell, A.D., Anthony, N.B., Pesti, G.M. and Aggrey, S.E. (2012) The effects of growth rate on leg morphology and tibia breaking strength, mineral density, mineral content, and bone ash in broilers. *Poultry Science* 91, 1790–1795.

Siegel, P.B., Honacker, C.F. and Rauw, W.M. (2009) Selection for high production in poultry. In: Rauw, W.M. (ed.) *Resource Allocation Theory Applied to Farm Animal Production*. CAB International, Wallingford, UK, pp. 230–242.

Sim, S. (2006) *Empires of Belief. Why We Need More Scepticism and Doubt in the Twenty-First Century.* Edinburgh University, Edinburgh, UK.

Simmons, I.G. (1995) Nature, culture and history. 'All just supply, and all relation'. In: Cooper, D.E. and Palmer, J.A. (eds) *Just Environments. Intergenerational, International and Interspecies Issues.* Routledge, London, UK, pp. 59–71.

Sinclair, U. (1906) *The Jungle.* Viking (1946 edn), London, UK.

Singer, P. (1975) *Animal Liberation*, 2nd edn, 1990. Thorsons, London, UK.

Singh, K. (2005) *Questioning Globalization.* Zed, London, UK.

Small, B., Murphy-McIntosh, A., Waters, W., Tarbotton, I. and Botha, N. (2005) Pastoral farmer goals and intensification strategies. Paper presented at the NZARES Conference, Nelson, NZ. Available at: https://www.researchgate.net/profile/Neels_Botha/publication/228429455_Pastoral_farmer_goals_and_intensification_strategies/links/5546f0160cf234bdb21daec7.pdf (accessed 31 March 2018).

Smart, J.A. (2004) Sudden illness and death in housed Merinos. In: *Proceedings of the Society of Sheep and Beef Cattle Veterinarians of the New Zealand Veterinary Association Annual Seminar 2004*, pp. 217–220. The Society of Sheep and Beef Cattle Veterinarians of the New Zealand Veterinary Association, Wellington, NZ.

Smiley, J. (1991) *A Thousand Acres.* Alfred Knopf, New York, NY.

Snow, C.P. (1959) *The Two Cultures.* Cambridge University Press, Cambridge, UK.

Somerville, M. (2006) *The Ethical Imagination. Journeys of the Human Spirit.* Melbourne University Press, Carlton, Australia.

Spedding, C. (2000) *Animal Welfare.* Earthscan, London, UK.

Speth, J.D. (2006) Housekeeping, Neandertal-style: hearth placement and midden formation in Kebara Cave (Israel). In: Hovers, E. and Kuhn, S. (eds) *Transitions before the Transition. Evolution and Stability in the Middle Paleolithic and Middle Stone Age.* Springer-Verlag, New York, NY, pp. 171–188.

Spooner, J.M., Schuppli, C.A. and Fraser, D. (2014) Attitudes of Canadian citizens toward farm animal welfare: a qualitative study. *Livestock Science* 163, 150–158.

Stafford, K. (2006) *The Welfare of Dogs.* Springer, Dordrecht, The Netherlands.

Stafleu, F.R., Grommers, F.J. and Vorstenbosch, J. (1996) Animal welfare: evolution and erosion of a moral concept. *Animal Welfare* 5, 225–234.

Stapledon, O. (1930) *Last and First Men.* Victor Gollancz, London, UK.

Steinbeck, J. (1939) *The Grapes of Wrath.* Penguin Books, Harmondsworth, UK.

Steinfeld, H., Wassenaar, T. and Jutzi, S. (2006) Livestock production systems in developing countries: status, drivers, trends. *Revue Scientifique et Technique de l'Office International des Épizooties* 25, 505–516.

Sterchi, B. (2018) *Cow.* Head of Zeus, London, UK.

Stolba, A. and Wood-Gush, D.G.M. (1984) The identification of behavioural key features and their incorporation into a housing design for pigs. *Annales de Recherches Veterinaires* 15, 287–298.

Strange, M. (1988) *Family Farming. A New Economic Vision.* University of Nebraska Press, Lincoln, NE.

Sullivan, R., Amos, N. and van de Weerd, H.A. (2017) Corporate reporting on farm animal welfare: an evaluation of global food companies' discourse and disclosures on farm animal welfare. *Animals* 7, 17.

Sunstein, C.R. and Nussbaum, M.C. (2004) *Animal Rights. Current Debates and New Directions*. Oxford University Press, Oxford, UK.

Surowiecki, J. (2004) *The Wisdom of the Crowds. Why the Many Are Smarter than the Few*. Abacus, London, UK.

Suzuki, D. and Taylor, D.R. (2009) *The Big Picture. Reflections on Science, Humanity and a Quickly Changing Planet*. Allen & Unwin, Vancouver, Canada.

Swift, J. (1726) *Gulliver's Travels*. Basil Blackwell edn (1959), Oxford, UK.

Swormink, B.K. (2017) Chicken of tomorrow is here today. *Poultry World*, 13 March. Available at: www.poultryworld.net/Meat/Articles/2017/3/Chicken-of-Tomorrow-is-here-today-103092E/ (accessed 6 June 2017).

Synovate Ltd (2011) *What New Zealanders Really Think about Animal Welfare*. MAF, Wellington, NZ.

Szabo, S., Tache, Y. and Somogyi, A. (2012) The legacy of Hans Selye and the origins of stress research: a retrospective 75 years after his landmark brief "Letter" to the Editor of *Nature*. *Stress* 15, 472–478.

Tannenbaum, J. (1991) Ethics and animal welfare: the inextricable connection. *Journal of the American Veterinary Medical Association* 198, 1360–1376.

Tauger, M.B. (2011) *Agriculture in World History*. Routledge, London, UK.

Teletchea, F. and Fontaine, P. (2014) Levels of domestication in fish: implications for the sustainable future of agriculture. *Fish and Fisheries* 15, 181–195.

Telkänranta, H., Marchant-Forde, J.N. and Valro, A. (2016) Tear staining in pigs: a potential tool for welfare assessment on commercial farms. *Animal* 10, 318–325.

Temple, D., Dalmau, A., Ruiz de la Torre, J.L. and Manteca, X. (2011) Application of the Welfare Quality® protocol to assess growing pigs kept under intensive conditions in Spain. *Journal of Veterinary Behavior* 6, 138–149.

Te Velde, H., Aarts, N. and Van Woerkum, C. (2002) Dealing with ambivalence: farmers' and consumers' perceptions of animal welfare in livestock breeding. *Journal of Agricultural and Environmental Ethics* 15, 203–219.

Thirsk, J. (1997) *Alternative Agriculture. A History from the Black Death to the Present Day*. Oxford University Press, Oxford, UK.

Thomis, M.I. (1970) *The Luddites. Machine-breaking in Regency England*. David & Charles, Newton Abbot, UK.

Thompson, P.B. (1995) *The Spirit of the Soil. Agriculture and Environmental Ethics*. Routledge, London, UK.

Thompson, P.B. (1999a) From a philosopher's perspective, how should animal scientists meet the challenge of contentious issues? *Journal of Animal Science* 77, 372–377.

Thompson, P.B. (1999b) Tying it all together. In: Hardy, R.W.E. and Segelken, J.B. (eds) *Agricultural Biotechnology: Novel Products and New Partnerships, NABC Report 8*. National Agricultural Biotechnology Council, Ithaca, NY, pp. 13–22.

Thompson, P.B. (2007) Ethics on the frontiers of livestock science. In: Swain, D.,

Chalmley, E., Steel, J.W. and Coffey, S.G. (eds) *Redesigning Animal Agriculture. The Challenge of the 21st Century.* CAB International, Wallingford, UK, pp. 30–45.

Thompson, P., Harris, C., Holt, D. and Pajor, E.A. (2007) Livestock welfare product claims: the emerging social context. *Journal of Animal Science* 85, 2354–2360.

Thompson, P. (2008) *The Ethics of Intensification. Agricultural Development and Cultural Change.* Springer, Dordrecht, the Netherlands.

Thomson, B. (2009) Beef industry bleeding badly and needs more than a bandage. *Heartland Beef*, May, pp. 12–13.

Thorp, B.H. (1994) Skeletal disorders in the fowl: a review. *Avian Pathology* 23, 203–236.

Thorpe, W.H. (1965) The assessment of pain and distress in animals. In: *Report of the Technical Committee to Enquire into the Welfare of Animals Kept under Intensive Livestock Husbandry Systems.* Her Majesty's Stationery Office, London, UK, pp. 71–79.

Thu, K.M. and Durrenberger, E.P. (eds) (1998) *Pigs, Profits, and Rural Communities.* State University of New York Press, Albany, NY.

Trafford, S. and Tipples, R. (2012) A foreign solution: the employment of short term migrant dairy workers on New Zealand dairy farms. Lincoln University, Christchurch, NZ. Available at: http://www.agrione.ac.nz/system/files/resource_downloads/Migrant%20Dairy%20Workers%20Literature%20Review_0.pdf (accessed 5 April 2018).

Tranchard, S. (2016) New ISO specification for better management of animal welfare worldwide. Available at: https://www.iso.org/news/2016/12/Ref2147.html (accessed 5 April 2018).

Trebilcock, K. (2017) Dairy 101: little bobby's many end uses. *NZ Farm Life Media.* Available at: https://nzfarmlife.co.nz/dairy-101-little-bobbys-many-end-uses/ (accessed 5 April 2018).

Tryon, T. (undated) *The Country-Man's Companion: OR, A New Method Of Ordering Horses & Sheep So as to preserve them both From Diseases and Causalties, Or, To Recover them if fallen Ill, And also to render them much more Serviceable and Useful to their Owners, than has yet been discovered, known or practised. And particularly to preserve Sheep from that Monsterous, Mortifying Distemper, The Rot.* Andrew Sowle, Shoreditch, UK. Available at: https://quod.lib.umich.edu/e/eebo/A63788.0001.001?rgn=main;view=fulltext (accessed 25 March 2018).

Tsovel, A. (2006) The untold story of a chicken and the missing knowledge in interspecific ethics. *Science in Context* 19, 237–267.

Tsushima, M. (2007) *An Outline of the Act on Welfare and Management of Animals.* Animal Welfare and Management Office, Ministry of the Environment, Tokyo, Japan. Available at: https://www.env.go.jp/nature/dobutsu/aigo/2_data/pamph/1912en/pdf/full.pdf (accessed 5 April 2018).

Tudge, C. (1998) *Neanderthals, Bandits and Farmers. How Agriculture Really Began.* Yale University Press, New Haven, CT.

Tudge, C. (2016) *Six Steps Back to the Land. Why We Need Small Mixed Farms and Millions More Farmers.* Green, Cambridge, UK.

Tulloch, G. (2011) Animal ethics: the capabilities approach. *Animal Welfare* 20, 3–10.

Twain, M. (1884) *Adventures of Huckleberry Finn. (Tom Sawyer's Comrade)*. Chatto & Windus, London, UK.

Uni, Z., Ferket, P.R., Tako, E. and Kedar, O. (2005) In ovo feeding improves energy status of late term chicken embryos. *Poultry Science* 84, 764–770.

Van Putten, G. (1988) Farming beyond the ability for pigs to adapt. *Applied Animal Behaviour Science* 20, 63–71.

Van Vliet, J., de Groot, H.L.F., Rietveld, P. and Verburg, P.H. (2015) Manifestations and underlying drivers of agricultural land use change in Europe. *Landscape and Urban Planning* 133, 24–36.

van de Weerd, H. and Sandilands, V. (2008) Bringing the issue of animal welfare to the public: a biography of Ruth Harrison (1920–2000). *Applied Animal Behaviour Science* 113, 404–410.

Veale, E.M. (1957) The rabbit in England. *Agricultural History Review* 5, 85–90.

Ventura, B.A., von Keyserlingk, M.A.G., Wittman, H. and Weary, D.M. (2016) What difference does a visit make? Changes in animal welfare perceptions after interested citizens tour a dairy farm. *PLoS ONE* 11, 5.

Verbeke, W. (2009) Stakeholder, citizen and consumer interests in farm animal welfare. *Animal Welfare* 18, 325–333.

Vialles, N. (1994) *Animal to Edible*. Cambridge University Press, Cambridge, UK.

Vila, C., Savolainen, P., Maldonado, J.E., Amorim, I.R., Rice, J.E., Honeycutt, R.L., Crandall, K.A., Lundeberg, J. and Wayne, R.K. (1997) Multiple and ancient origins of the domestic dog. *Science* 276, 1687–1689.

Villalba, J.J. (2016) *Animal Welfare in Extensive Production Systems*. 5M Publishing, Sheffield, UK.

Vines, G. (1994) How far should we go? *New Scientist* 141 (1912), 12–13.

Walsh, S.W., Williams, E.J. and Evans, A.C.O. (2011) A review of the causes of poor fertility in high milk producing dairy cows. *Animal Reproduction Science* 123, 127–138.

Wangerin, W. Jr (1978) *The Book of the Dun Cow*. Penguin Books, Harmondsworth, UK.

Ward, S. (2007) Fewer but bigger dairy farms predicted. *The New Zealand Herald*, 20 March, p. 13.

Waterhouse, A. (1996) Animal welfare and sustainability of production under extensive conditions—a European perspective. *Applied Animal Behaviour Science* 49, 29–40.

Wathes, C.M., Corr, S.A., May, S.A., McCulloch, S.P. and Whiting, M.C. (2013) *Veterinary & Animal Ethics. Proceedings of the First International Conference on Veterinary and Animal Ethics, September 2011*. Wiley-Blackwell, Oxford, UK.

Weary, D.M., Ventura, B.A. and von Keyserlingk, M.A.G. (2016) Societal views and animal welfare science: understanding why the modified cage may fail and other stories. *Animal* 10, 309–317.

Webster, A.J.F. (1992) Energy expenditure: studies with animals. In: Widdowson,

E.M. and Mathers, J.C. (eds) *The Contribution of Nutrition to Human and Animal Health*. Cambridge University Press, Cambridge, UK, pp. 23–32.

Webster, A.J.F. (2009) The Virtuous Bicycle: a delivery vehicle for improved farm animal welfare. *Animal Welfare* 18, 141–147.

Webster, J. (1994) *Animal Welfare. A Cool Eye towards Eden*. Blackwell Science, Oxford, UK.

Webster, J. (2005a) *Animal Welfare: Limping towards Eden*. Blackwell, Oxford, UK.

Webster, J. (2005b) The assessment and implementation of animal welfare: theory into practice. *Revue Scientifique et Technique de l'Office International des Épizooties* 24, 723–734.

Wechsler, B. (1996) Rearing pigs in species-specific family groups. *Animal Welfare* 5, 25–35.

Wells, R.C. (1997) Mr William Cobbett, Captain Swing, and King William IV. *Agricultural History Review* 45, 34–48.

Wemelsfelder, F., Hunter, T.E., Mendl, M.T. and Lawrence, A.B. (2001) Assessing the 'whole animal': a free choice profiling approach. *Animal Behaviour* 62, 209–220.

Wertheim, M. (1999) *The Pearly Gates of Cyberspace: A History of Space from Dante to the Internet*. Doubleday, Milsons Point, Australia.

Westerfield, R.B. (1915) *Middlemen in English Business. Particularly between 1660 and 1760*. Yale University Press, New Haven, CT (David & Charles Reprint, Newton Abbot, UK, 1968).

Whiting, T.L. and Marion, C.R. (2011) Perpetration-induced traumatic stress—a risk for veterinarians involved in the destruction of healthy animals. *Canadian Veterinary Journal* 52, 794–796.

Whyte, K.P. and Thompson, P.B. (2012) Ideas for how to take wicked problems seriously. *Journal of Agricultural and Environmental Ethics* 25, 441–445.

Wilkins, L.J., McKinsty, J.L., Avery, N.C., Knowles, T.G., Brown, S.N., Tarlton, J. and Nicol, C.J. (2011) Influence of housing system and design on bone strength and keel bone fractures in laying hens. *The Veterinary Record* 169, 411–412.

Williscroft, C. (2007) *A Lasting Legacy. William Davidson 125. A 125-year History of New Zealand Farming since the First Frozen Shipment*. NZ Rural Press, Auckland, NZ.

Wilson, E.O. (2001) *The Diversity of Life*. Penguin, London, UK.

Wilson, E.O. (2014) *The Meaning of Human Existence*. Liveright, New York, NY.

Wilson, E.O. (2016) *Half-Earth. Our Planet's Fight for Life*. Liveright, New York, NY.

Wilson, J. and Tipples, R. (2008) Employment trends in dairy farming in New Zealand 1991-2006. Lincoln University, Christchurch, NZ. Available at: http://researcharchive.lincoln.ac.nz/bitstream/handle/10182/574/agls_rr_2.pdf?sequence=3 (accessed 7 April 2018).

Wilson, P.R. (2002) Advances in health and welfare of farmed deer in New Zealand. *New Zealand Veterinary Journal* 50 Suppl, 105–109.

Winter, M., Fry, C. and Carruthers, S.P. (1998) European agricultural policy and farm animal welfare. *Food Policy* 23, 305–323.

Wood-Gush, D.G.M. (1959) A history of the domestic chicken from antiquity to the 19th century. *Poultry Science* 38, 321–326.

Wood-Gush, D.G.M. and Beilharz, R.G. (1983) The enrichment of a bare environment for animals in confined conditions. *Applied Animal Ethology* 10, 209–217.

Wood-Gush, D.G.M. and Vestergaard, K. (1989) Exploratory behavior and the welfare of intensively kept animals. *Journal of Agricultural Ethics* 2, 161–169.

Wood-Gush, D.G.M. and Vestergaard, K. (1991) The seeking of novelty and its relation to play. *Animal Behaviour* 42, 599–606.

Woods, A. (2011a) The history of veterinary ethics in Britain, ca. 1870–2000. In: Wathes, C.M., Corr, S.A., May, S.A., McCulloch, S.P. and Whiting, M.C. (eds) *Veterinary & Animal Ethics. Proceedings of the First International Conference on Veterinary and Animal Ethics, September 2011.* Wiley-Blackwell, Oxford, UK, pp. 3–18.

Woods, A. (2011b) From cruelty to welfare: the emergence of farm animal welfare in Britain, 1964–71. *Endeavour* 36, 14–22.

Wright, R. (1994) *The Moral Animal. Evolutionary Psychology and Everyday Life.* Abacus, London, UK.

Wroblewski, D. (2008) *The Story of Edgar Sawtelle.* Harper Collins, New York, NY.

Wykes, D.L. (2004) Robert Bakewell (1725–1795) of Dishley: farmer and livestock improver. *Agricultural History Review* 52, 38–55.

Yeates, J.W. (2011) Is 'a life worth living' a concept worth having? *Animal Welfare* 20, 397–406.

Yeates, J.W., Röcklinsberg, H. and Gjerris, M. (2011) Is welfare all that matters? A discussion of what should be included in policy-making regarding animals. *Animal Welfare* 20, 423–432.

Zeder, M.A. (2006) Central questions in the domestication of plants and animals. *Evolutionary Anthropology* 15, 105–117.

Zeuner, F.E. (1963) *A History of Domesticated Animals.* Hutchinson, London, UK.

Zimdahl, R.L. (2002) Moral confidence in agriculture. *American Journal of Alternative Agriculture* 17, 44–53.

Zimdahl, R.L. (2006) *Agriculture's Ethical Horizon.* Academic Press, Burlington, MA.

Index

Aaltola, E. 165
An Act to Prevent the Cruel and Improper Treatment of Cattle (1822) 128
Adams, Douglas 204
   The Restaurant at the End of the Universe 77
Adams, Richard, *Watership Down* 7
advocacy groups 137–40
Aesop, *Goose and the Golden Egg* 207
agribusiness 71–5, 88, 108, 109, 195, 222, 223
Agricultural Revolution 46, 238
   Middle Ages 61–6
   Neolithic 53–61
agriculture 7–8
   as business and way of life 31
   and culture 214–19, 222
   enclosures 65
   failures in 31–2
   goals 219–20
   good 234–7
   modern changes in 75–81
   open-field systems 61–3
   reduce consumption 234
   reward farmers 234
   stages of development 46–7
   technological innovation 63
Agriculture (Miscellaneous Provisions) Act (1968) 130–1
American Society for the Prevention of Cruelty to Animals 121
Ancient Egypt 123–4
Animal Liberation Front 139
animal rights 132–3, 139, 160–1
animal welfare 14–18, 38, 39
   ability to adapt 25–6
   assessing/auditing 150, 152
   assurance schemes 149–53
   burden of 205–10
   challenges 30
   charities/advocacy groups 137–9
   complexities of 33–4
   consumer attitudes 153–9
   definitions/understanding 19–27
   delivering good practices 170–2
   disconnect between realities/expectations 34
   economics of 165–70
   effect of extensive farming on 91–4
   environment 22–4, 25–6
   equitable distribution of costs/benefits 159
   fitting people to animals/farms 239–42
   five freedoms 135–6
   and good citizenship 234–7
   health/disease 140–3
   humanitarian concerns 133–59
   impact of factory farming 143–4
   and increased productivity 105–6
   international standards 136–7
   legislation 158
   linked with food safety 104
   market/private-based approaches 158
   and modern culture 216–19
   as moral problem 220–1
   poor 31–4, 84, 116, 217–18, 244
   responsibility for 14–16, 196, 240
   as shared obligation 235–6
   as social construct 170–1, 243
   social ethics concerning 134–40, 159–65
   social expectations 27–9

animal welfare (cont.)
  social/people dimension 26–7
  subjective judgements 29–30
  suggested principles 232–3
  as symptom/reflection of human well-being 215
  terminology 20
  understanding natural behaviour 143–8
  as wicked 30–4, 239
  see also husbandry
Animal Welfare Board of India 135
animal welfare science 140–9
animal-human relationship
  affluent contract 205–10
  African saying 174
  animal products 178–9
  animals as source of food 178
  animals as source of labour 177–8
  in art 181, 182, 183
  benefits of 181
  burden of animal welfare 205–10
  cattle and gauchos 203–4
  changing 198–205, 224, 238–42
  characteristics of a good life 208
  disconnection between 223–8, 244–5
  diversity of 175–6
  as fair 246–7
  Haitian proverb 186
  historical aspects 183–5
  importance of animals 174
  killing animals 189–93, 227
  modern connections 228–37
  modern disconnects 195–8
  myths and stories 185–6
  origins of disconnect 194–5
  pests and diseases 181
  pets and companions 181
  and rabbits 198–201
  sheep and shepherds 201–3
  social/ecological interdependence 185
  sports, recreation, entertainment 179–80, 181
  symbolic 181, 183
  visibility issues 185–9
  Yoruba proverb 195
animals
  body temperature 23
  breeding/selection 93–4, 98, 100–1
  castration 17, 93, 94

  costs/benefits of using 38
  cruelty/ill-treatment of 16–17, 126–32
  culling 10, 13, 42–3, 58, 65, 100, 103, 191
  economics of 165–70
  and farming 13–16, 17–18
  five freedoms 14, 18
  habitat 13
  killing 189–93, 227
  management practices 87–8
  neglect/abuse 18–19
  physiological/behavioural indicators 22–6
  relationship with 35–6
  stressors 27, 104, 144
  suckling 27–9
  suffering of 13–19, 35–6, 129, 131
  tail docking 17, 94
  thwarted preferences/discomfort 12–13
  understanding behaviour of 143–8
  well-being of 130–1
  wild 53
anthropomorphism 48, 132, 143–4, 166, 181, 183
Aquatic Health Code 136
Arvesen, Ane-Karine 33
Asoka (or Ashoka the Great) 124
assurance programmes 17, 107, 115, 118, 119, 149–53, 224, 240
Atwood, M. 39
Auden, W.H. 213
Australopithecines 5

badgers 9–10, 11, 181
Bagshot (Surrey) 121
Bakewell, Robert 38, 64
Banksy, Trolley Hunters 245–6
Banner, Michael 136
Bennett, John 121
Benson, Richard, The Farm 118, 119–20, 122
Bentham, Jeremy 127
Benton, T., Natural Relations: Ecology, Animal Rights and Social Justice 177
Berry, Wendell 223
  The Unsettling of America 195, 222
  'Whose Head is Using the Farmer and Whose Head Is the Farmer Using?' 219

bete-machine 126
the Bible 124
bioethics 236–7
biology 37
biophilia 205
Black Death 63
Black Tatars (Siberia) 4
Body, R., *Farming in the Clouds* 88
Book of the Dead (or *Spells for Going Forth by Day*) 123–4
The Book of the Dun Cow 6
Brambell, F.W. Rogers 130
Brambell Report (1965) 130, 131, 132, 134, 135
British Society of Animal Production, *Farm Animal Welfare - Who Writes The Rules?* 155
Brophy, Brigid, *The Rights of Animals* 132
Browne, W.P., *Sacred Cows and Hot Potatoes* 164
Burke, Edmund 215
Business Benchmark on Farm Animal Welfare 139

Cape ground squirrels 22
Captain Swing riots (1830) 66, 83
Carson, Rachel, *Silent Spring* 12
cattle 6, 91, 127
  grass-fed systems 138
  health of 102–3
  humane handling of 148
  killing surplus calves 220
  Middle Ages revolution 61
  Neolithic domestication 53, 55, 57, 60
  retailer/consumer demands 71
charities 137, 139, 188
Cheyenne 184
children 29, 37, 76, 129, 131, 181, 183, 185–6, 189, 223–4
Church Hole Cave, Cresswell Crags (UK) 43
Churchill, Winston 149
circus 180
citizenship 234–7
Colorado 8, 30
Columella, Lucius Junius Moderatus, *On Agriculture* (*De Re Rustica*) 124–5
commerce 68–9

connection mechanisms
  and animal welfare system 229–30
  changing animal/human relationship 224
  changing perception of animals 227–8
  changing the system 224–5
  consumer/farm linkage 224
  decision-making 230–1
  images of animals 230
  importance of reconnection 223–4
  insights from other cultures 225–7
  learn about/manage humanity's collective behaviour 225
  modern sophisticated connections 228–37
  roles/responsibilities of society 231–2
  understanding the system 228–9
corporate social responsibility 224
coyotes 8, 10, 11, 30
Crump, Barry, *A Good Keen Man* 43
A Crusade Against All Cruelty to Animals 129
culture 214–19, 220

dairy industry 6–7, 20, 84, 97, 99, 100, 106–7, 116–18, 138–9, 148, 204
Daley, David 33
Darwin, Charles 3
  On the Origin of the Species 53
Davis, William 121
Dawkins, Marian Stamp, *Animal Suffering: The Science of Animal Welfare* 131
deer 38
  domestication 45–6, 60
  farming of 43–4
  fence-pacing 45
  hunting and managing 42–4, 180, 212
  introduction 42
  selection/reproduction 44–5
The Deer Menace conference (1930) 43
Descartes, René 125–6, 129
Dick, Philip K., *Do Androids Dream of Electric Sheep?* 12
disconnection with animals *see* connection mechanisms
disease 9, 10, 13, 25, 32, 44, 56, 58, 59, 72, 73, 75, 89, 92, 94, 100–5, 178, 181
dog-wolves 2–3
dogs 40, 53, 55, 180, 191
  myths concerning 4

dogs (*cont.*)
  relationship with humans 1–6, 177
  statue of 1, 42
  treatment of 3–4
domestication
  animal selection 54–5
  animal-centred/zoological
    understanding of 56
  compared with wild animals 58
  costs/benefits 59–61
  effects of 218–19
  as evolutionary strategy 53–4
  human-centred, sociocultural/
    anthropological criteria 56–7
  impact on environment 59–60
  main centres of farming 55
  morphological, physiological,
    behavioural traits 57–9
  need for social organization/technology
    57
  Neolithic revolution 53–61
  species of animals 55
donkeys 6, 55
Durrenberger, E.P. (with K.M. Thu), *Pigs, Profits and Rural Communities* 73, 108

ecology 225
  availability of resources 85, 88
  foraging/feeding 85, 87
  intensification of farming 85, 87–9
  overcoming constraints 85–9
  sustainable 217–19
Edinburgh Pig Park 144
environment 35
  and animal welfare 91–4
  modification/land-use 89–91
environmental ethics 212–13, 214
environmental resilience 39
Ernst, Max, *City with Animals* 243, *244*
Erskine, Thomas 127–8
ethics 134–40, 159–60, 222
  animal rights 160–1
  common sense/tradition 160
  economic (cost-benefit analysis) 166–70
  environmental 212–13, 214
  good 221
  medical 161
  philosophy of morality 160
  pragmatism 162

  relationships 161–2
  theories 160–3
  utilitarianism 160
  virtue 162
EU Welfare Quality scheme 150, 151
extensive farming systems 89–94

factory-farming 129–30, 143–4, 222
Family Pen System 144–6
Farm Animal Welfare Council/Committee 16, 39, 130, 135, 192, 210, 224
  Opinion on the Welfare of the Dairy Cow 116–18
farm animals
  body temperature 22–3
  children and 223–4
  connecting with 223–5
  as economic resource 221–2
  effect of high productivity/
    intensification 98–104
  environment 23–4
  health of 100–4
  human interaction with 5, 188–9
  improvements in 105–6
farm workers 12
  immigrant labour 75
  social/environmental stressors 107–9
  treatment of 31
farm-gate returns 74
farmers 35
  and animal welfare 14–16, 196
  health of 14–15
  impact of high farming/hard work on 106–18
  numbers of 70, 76
  relationship with groups outside farming 80–1
  social/environmental stressors 107–9
farming 7–8
  adoption of 52–3
  business focus 10–11
  choosing/developing production systems 71
  compromise in 10
  conventional image 204
  criticisms of 11–12, 33
  as cultural practice 195
  genetically modified image 204
  good 219–28

husbandry procedures 17–18, 24
increasing cost of 74
initiatives/solutions 215–16
internal/external relations 11
job satisfaction in 75
landholding/social relationships 65–6
mechanization 70–1
Middle Ages revolution 61–6
modern changes in 69–81
as nexus of two worlds 80
origins 50–1
reliance on multitude of industries 71
romantic/historical image 204
rural unrest 66
as shared obligation 235–6
size of farms 69–70, 83–4
soil maintenance 115
structural factors 87
surplus, wealth, modern elite 75–81
as sustainable 112–13
technological innovations 63, 74–5
tensions between labourers/owners 66
traditional portrayals/new perceptions 164
Farming Community Network Helpline 16
Finney, T. 193
firestick farming 50
fish 6, 98, 179, 180
Five Freedoms 14, 18, 135
food
  and the Agricultural Revolution 61–6
  and animal welfare 154–6
  animals as source of 2, 3, 6, 27, 29, 38, 178, 189–94
  children's knowledge of 188
  cost of production 114–15
  efficiency of production 207, 209
  and hunter-gatherers 47–53
  increased production 97–8
  markets and middlemen 66–75
  modern technology/production 75–81, 83
  and Neolithic Revolution 53–61
  origins of fresh foods in a supermarket 79
  price/availability 35, 78, 79, 80
  production ethics 149–50, 219–20
  profitable, sustainable, safe production of 8, 11
  safety linked with animal welfare 104
  surplus production 76
Food Ethics Council, *Farming Animals for Food: Towards a Moral Menu* (2001) 138
Ford, Henry 70
Fox, Michael, *Superpigs and Wondercorn* 12
Friedman, Thomas, *The Lexus and the Olive Tree* 229
Fukuoka, Masanobu, *The One-Straw Revolution* 44

geese 66–7
Gell-Mann, Murray, *The Quark and the Jaguar* 229
Gérome, Jean-Léon 230
Glover, Denis, *The Magpies* 114–15
goats 53, 55, 57, 60, 91, 94, 191
Goethe, Johann Wolfgang von 126
Goldschmidt, W., *As You Sow* 108, 109
Good Agricultural Practices (UN) 138
Goodall, Jane 247
Grahame, Kenneth, *The Wind in the Willows* 9
Gravettian culture 50–1
Great Depression 12, 71, 220

hares 42, 180, 199
Harrison, Ruth 18, 165, 195, 196, 197–8
  Animal Machines 12, 129–30, 131, 134, 135, 222
Henry VIII 157
Herod the Great 153
Herzen, Aleksandr 215
high farming
  background 83–5
  ecological constraints 85–9
  extensive/intensive 89–94
  hard work/undesirable traits 98–104
  impacts on farmers 106–18
  improvements in farm animals 105–6
  performance 96–8
  reflections on modern production 118–20
  stockmanship 94–6
Hindus 124
homeostatic mechanisms 22–6
horses 6, 12, 51, 57, 126, 180
Humane Society of the United States 121
humanitarian concerns *see* animal welfare
Humboldt, Alexander von 228

Hume, David 126
hunter-gatherers 2–3, 7, 36, 46
  adoption of farming 52–3
  change to farming 48
  changes in the mind 48–9
  development of material culture 50–1
  fall and rise of 243–8
  guided by myths and legends 123
  nomadic lifestyle 47–8
  organization of 52
  specialization/division of tasks 49–50
  view of nature/natural world 48
husbandry 39
  in antiquity 123–5
  background 122–3
  delivering good animal welfare 170–2
  effects of intensification 218
  in the Enlightenment 126–7
  ethics and economics 159–70
  modern humanitarian concerns 133–59
  in the nineteenth century 127–8
  procedures 17–18, 23, 36, 37
  provision of water 121–2
  and rabbits 198–201
  in the Renaissance 125–6
  and sheep 201–3
  species-specific behaviour 143–8
  in the twentieth century 128–33
  see also animal welfare

Industrial Revolution 46, 60, 238
intensive farming 80, 83, 84, 187, 243–4
  animal management practices 87
  animal welfare 38–9, 93–4
  costs/benefits 88–9
  and culture 220
  effects of 218–19
  extensive farming systems 91–4
  farm structural factors 87
  harms related to high productivity 98–104
  increase in 70–4
  increased food production 97–8
  land use 89–91
  modification of environment 89–90
  performance/costs of production 113
  and poor animal welfare 214–15
  problems 88–9
  proximate/ultimate 196–7
  reasons for persevering with 109–14
  risks of 217
  undesirable side-effects 105–6
International Organization for Standardization (ISO) 137
Islam 124, 191

Jack, Alex 118–19

Kant, Immanuel 127, 153
Kebara cave, Mt Carmel (Israel) 42
Kilgour, Ron 141, 146–8
Kirikongo 247
Koyukon Indians of Alaska 194

lactase 6–7
landless farming 90
Lapenotiere, Captain John Richards 121
Lawler, A. 247
Laxton (Nottinghamshire) 61
Le Guin, Ursula, *The Ones Who Walk Away from Omelas* 185–6, 207, 237, 240
LEAF *see* Linking Environment And Farming
Leakey, Louis 247
legislation 15, 17, 43–4, 107, 117, 134, 135, 149, 153, 157–8, 159, 170, 200, 210, 236
Leopold, Aldo 212, 213, 214, 216
Linking Environment And Farming (LEAF) 188
llamas 6, 55
Lockwood, J.A., *Grasshopper Dreaming* 196

McInerney, J. 241
markets 66–9
Martin, Richard 128
Martin's Act (1822) 128
Metropolitan Drinking Fountain and Cattle Trough Association 121
middlemen 67–9, 77–8
Midgley, Mary, *Animals and Why They Matter* 133
Mithen, S., *The Prehistory of the Human Mind* 48
mixed farming 90
More, Thomas, *Utopia* 192
myths and stories 123, 163–5, 185–6, 192, 194, 207

Nairne, Patrick 236–7
Natufians 51
Neanderthals 42, 49
Neolithic period 47, 52
  agricultural revolution 53–61
Neolithic Revolution 237–8
New Dishley (or New Leicester) sheep 64–5
Newkirk, Ingrid 139
Nile Valley 59–60
Nottingham Goose Fair 66–7
Nuffield Council on Bioethics 236
National Animal Welfare Advisory Committee 135

Observer newspaper 130
OIE *see* World Organisation for Animal Health (OIE or Office International des Épizooties)
open-field system 61–3
Orwell, George, *Animal Farm* 52, 81
Ozeki, Ruth, *My Year of Meat* 12

Pacheco, Alex 139
Paleolithic period 46, 47
Palmer, Sir Geoffrey 237
pastoral farming 7, 60, 65, 85, 89, 90, 91, 96, 124, 165, 195, 202, 204, 205, 206, 228, 240
People for the Ethical Treatment of Animals (PETA) 139
PETA *see* People for the Ethical Treatment of Animals
Pew Commission, *Putting Meat on the Table: Industrial Farm Animal Production in America* (2008) 137–8
pigeons 55, 125, 181
pigs 12, 93, 99, 180
  breeding, confinement, intensification 73–4, 98, 138, 220
  health of 103–4
  human health related to 108–9
  Middle Ages revolution 61
  Neolithic domestication 55
  symbolic view of 181
  understanding behaviour of 144–6
Piper, H. Beam, *Little Fuzzy* 12
Popper, Sir Karl 240
Potter, Van Rensselaer 212–13, 216, 237

poultry 6, 12, 204
  egg production 56, 73, 78, 90, 98, 99
  feed-restricted birds 102
  health of 102
  intensification/specialization 71–3, 94, 138, 144, 220
  Neolithic domestication 53, 55, 57
  retail price of 78
Pratchett, Terry, *Where's My Cow?* 187–8
Preece, R., *Awe for the Tiger, Love for the Lamb* 184, 194
prehistory 47–9
production diseases 100–4
  genetic predispositions 101
  husbandry practices 102
  nutritional conditions 101
proto-farming 50–3, 89
Pythagoras 124
Pythagorean Diet 124

rabbits 6-10, 12, 13, 29, 42, 51, 83, 180, 181, 198-201
rats 6, 181
Rebanks, James, *The Shepherd's Life* 12
reconnection with animals *see* connection mechanisms
Regan, Tom, *The Case for Animal Rights* 132–3
reindeer 51, 60
Renaissance 126–7
Report of the Committee to Consider the Ethical Implications of Emerging Technologies in the Breeding of Farm Animals (1995) 136
Report of the Technical Committee to Enquire into the Welfare of Animals Kept under Intensive Livestock Husbandry Systems (1965) 130
Reynnells, R. 122
Robinson, Kim Stanley 2
Rollin, Bernard 133–4
Romanes, George 3
Royal Society for the Prevention of Cruelty to Animals (RSPCA) 130
Ryder, Richard 132

St Jerome 230, 231
Salt, Henry, *Animal Rights: Considered in Relation to Social Progress* 128
Saul, John Ralston, *On Equilibrium* 222–3
Schweitzer, Albert 128–9

Selye, Hans 141–3
  The Stress of Life 143
service industries 77
Sewell, Anna, *Black Beauty* 12, 135
sheep 89, 91, 126
  breeding 64–5, 94
  husbandry of 201–3
  Middle Ages revolution 61, 62
  Neolithic domestication 50, 53, 55, 57, 60
  symbolic view of 181, 183
Sinclair, Upton, *The Jungle* 135
Singer, Peter, *Animal Liberation* 132
Smiley, Jane, *A Thousand Acres* 12
Solomon, King 153
Somerville, Margaret 229
  The Ethical Imagination 220
South America 29, 203–4
speciesism 132
SS *Dunedin* 68
Stapledon, Olaf, *Last and First Men* 214
Steele, Celia 72
Steinbeck, John, *The Grapes of Wrath* 12
stockmanship 94–6, 100, 218
Stolba, Alex 144
Stone Age 42, 47, 49, 183, 229
Strange, M., *Family Farming: A New Economic Vision* 70
subsistence farming 68–9
Sunday Times 132
Surowiecki, J., *The Wisdom of Crowds* 236
sustainability 8, 11, 112–13, 217–19, 223
Suzuki, David 239

Talmud 124
Terrestrial Animal Health Code 136, 137
Thorpe, William H. 131, 140
Thu, K.M. (with E.P. Durrenberger), *Pigs, Profits and Rural Communities* 73, 108
Tolstoy, Leo, *What Then Must We Do?* 215
transport 68, 71, 73–4, 94
Tryon, Thomas, *The Country-Man's Companion* 126–7

Turnspete (or turnspit) dog 3
Tuvan people 227
Twain, Mark, *Adventures of Huckleberry Finn* 14
Tyson, John 71

United Nations Food and Agricultural Organization (UN FAO) 138
  Animal Legal Defense Fund 138

Venezuela 29
vets 15, 31, 32, 37, 58, 74, 88, 92, 103
Voltaire, François-Marie Arouet 126

ways of knowing 222–3
  common sense/shared knowledge 222
  ethics 222
  imagination 222
  memory/history 222
  reason 222
Webster, John 135
  Animal Welfare: Limping towards Eden 105
WEIRD *see* Western, Educated, Industrialized, Rich and Democratic
Western, Educated, Industrialized, Rich and Democratic (WEIRD) 225–8
Wilde, Oscar 228
Wilson, E.O., *Half-Earth* 34
wolves 2, 6, 92, 212
Wood-Gush, David 141, 143–4
woollen products 77
World Organisation for Animal Health (OIE or Office International des Épizooties) 136–7
Wroblewski, D., *The Story of Edgar Sawtelle* 246

Yarralin people (Australia) 4
'Young Carnivores Shouldn't Be so Chicken' (2011) 189

zoonotic diseases 104–5